Horizons Science

John Chapman

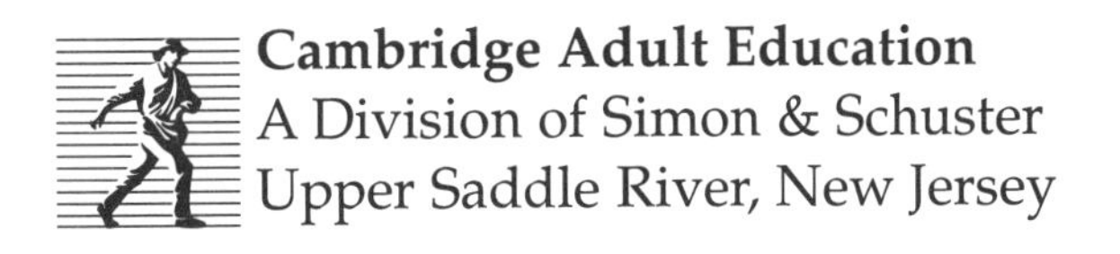
Cambridge Adult Education
A Division of Simon & Schuster
Upper Saddle River, New Jersey

Executive Director: Mark Moscowitz
Project Editors: Robert McIlwaine, Bernice Golden, Keisha Carter, Laura Baselice, Lynn Kloss
Writer: John Chapman
Series Editor: Roberta Mantus
Consultants/Reviewers: Marjorie Jacobs, Cecily Kramer Bodnar
Production Manager: Penny Gibson
Production Editor: Nicole Cypher
Marketing Manager: Will Jarred
Interior Electronic Design: Flanagan's Publishing Services, Inc.
Illustrator: Accurate Art, Inc. & Andre V. Malok
Photo Research: Jenifer Hixson
Electronic Page Production: Flanagan's Publishing Services, Inc.
Cover Design: Armando Baez

Photo Credits: p. 10: Courtesy of American Airlines; p. 11: Maine Dept. of Economic Development; p. 13: Union Pacific Railroad Photo; p. 42: Picture Cube; p. 43: Library of Congress; p. 62: UPI/Bettmann; p. 64: The Bettmann Archive; p. 68: The Bettmann Archive; p. 70: UPI/Bettmann

Printed in the United States of America.
1 2 3 4 5 6 7 8 9 10 99 98 97 96 95
ISBN: 0-8359-4630-4

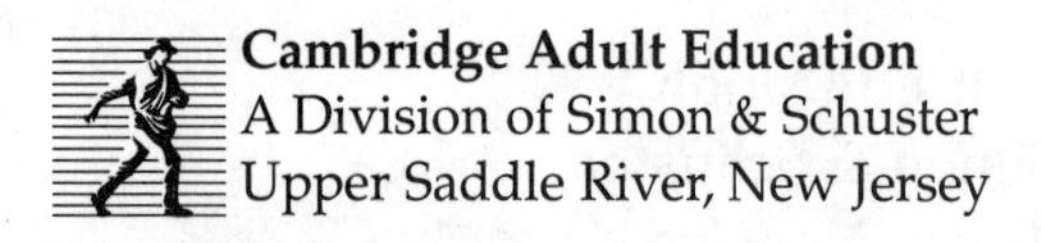

Contents

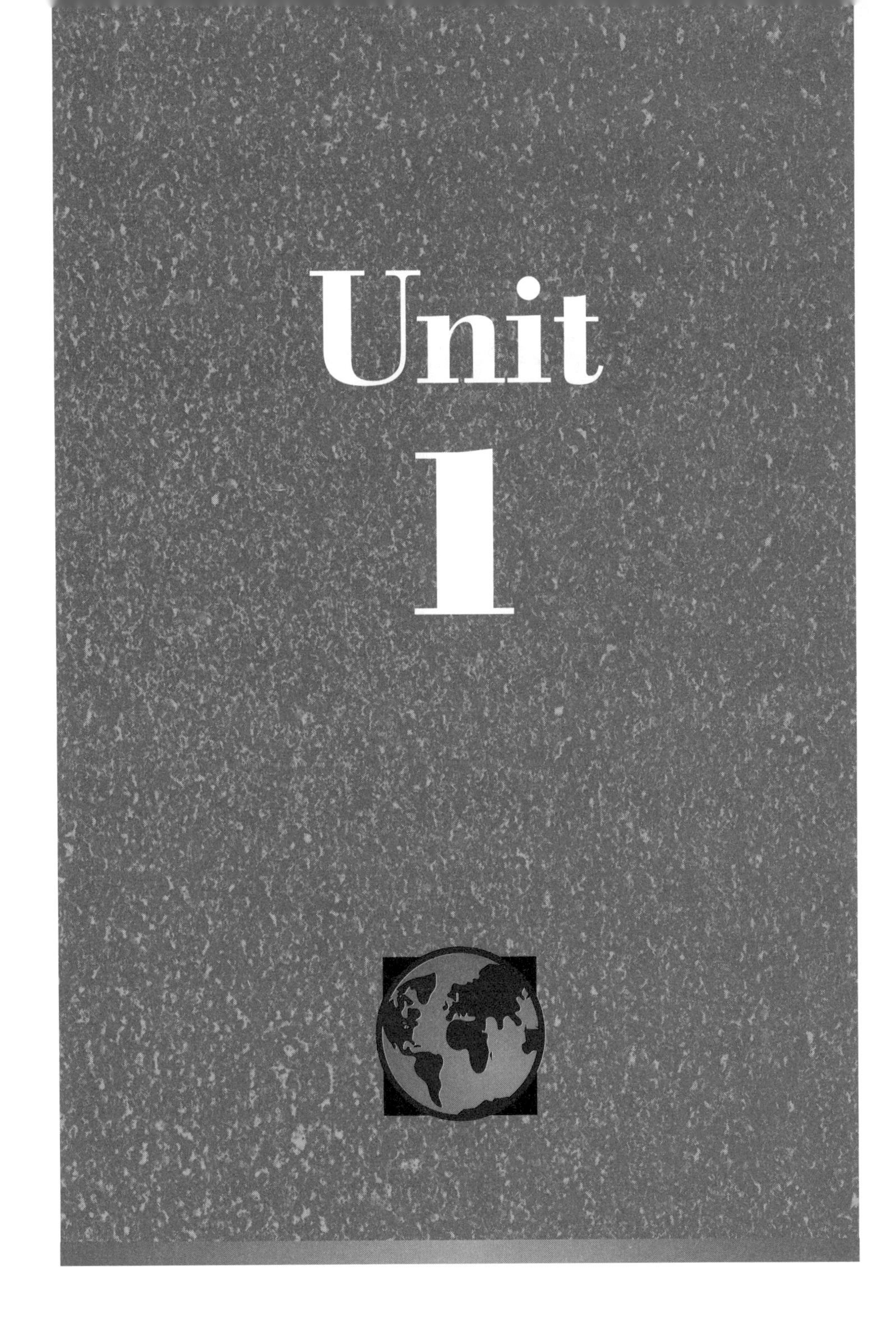

Earth Science

Lesson

1

The Beginning of the Earth

What You Know One good way of figuring out how to say words that are new to you is to say them like words that you already know. For example, you know the words *take* and *lake*. When you come to a word you don't know, like *slake*, you pronounce it like *take* and *lake*.

How It Works

There are five **vowels** (VOW-uhlz) in the alphabet. They are *a, e, i, o,* and *u*. You can say each vowel two different ways.

Say the words *bake* and *back*. The *a* in *bake* sounds just like the letter *a* in the alphabet. Vowels that are said the way they sound in the alphabet have **long vowel sounds**.

Here are some examples of words with long vowel sounds:

a as in *make*   *e* as in *be*   *i* as in *size*

o as in *close*   *u* as in *cube*

Try It

Read the following sentences. Then look at the underlined words. Some of them have long vowel sounds. Write only the words with long vowel sounds in the answer blanks.

1. Some people think the nine planets came from the sun. They say a star passed close to the sun. It made a piece of the sun break off.

__________ __________

__________ __________

Here are the long vowels: the *i* in *nine*, the *a* in *came*, the *o* in *close*, and the *a* in *made*. The vowels in *think, sun, star,* and *from* are not said the way they sound in the alphabet. They are not long vowels.

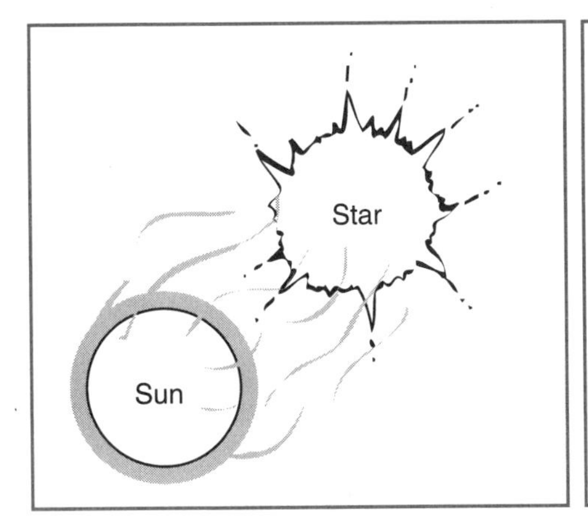

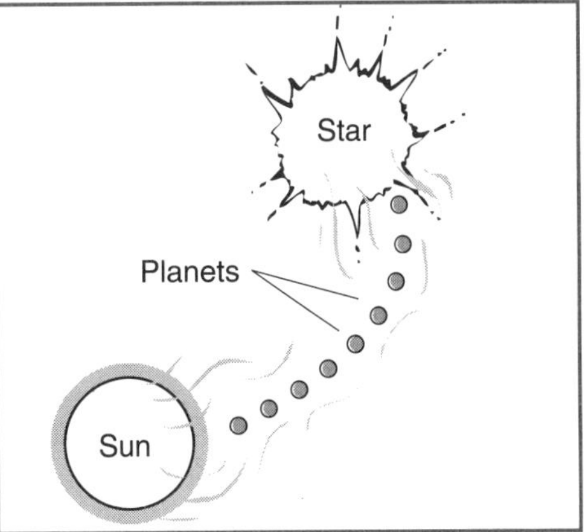

Some people believe that a star passed close to the sun.
It caused a piece of the sun to break off.

How It Works

Some vowels are not said the way they sound in the alphabet. In the word *hot,* the letter *o* doesn't sound like the letter *o* in the alphabet. Words that have these kinds of sounds have **short vowel sounds**.

Here are some more examples of words with short vowel sounds:

a as in *gas*   *e* as in *less*   *i* as in *hit*

o as in *pop*   *u* as in *dust*

Try It

Read the following sentences. Then look at the underlined words. Some of them have short vowel sounds. Write only the words with short vowel sounds in the answer blanks.

1. This big piece of the sun broke into lots of little pieces.

   ____________ ____________

   The *i* in *big* and the *o* in *lots* are short vowels. The *o* in *broke* is a long *o*.

2. These little pieces of the sun became the planets. There are nine of them. The earth is the fifth largest planet.

   ____________ ____________

   The *u* in the word *sun* and the *i* in *fifth* are short vowels. The *i* in *nine* is a long vowel.

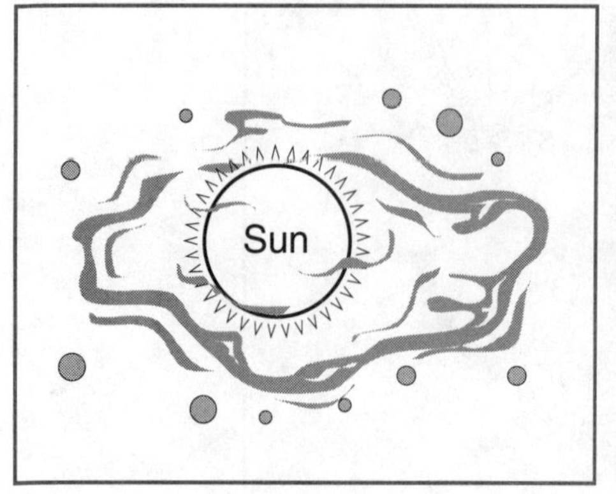

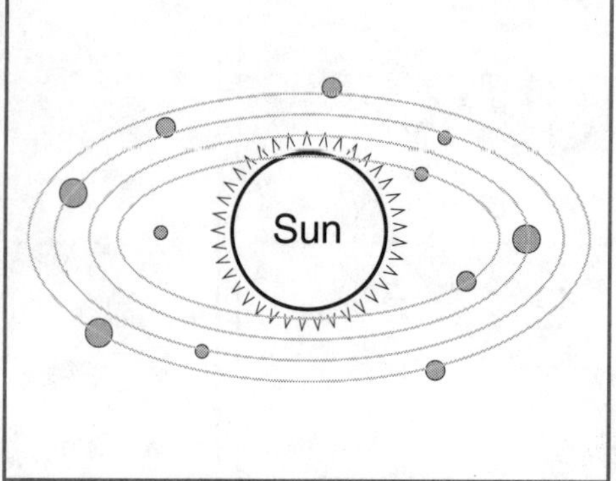

A cloud of dust and gas formed the sun. Then the planets broke off from the sun.

Practice

Write each underlined word on a long answer blank. Then write *Long* for words with long vowel sounds or *Short* for words with short vowel sounds in the short answer blank after each word.

1. Some scientists think that the sun began as space dust and gas.

   ____________ ______ ____________ ______

2. At first, this dust and gas looked like a cloud.

   ____________ ______ ____________ ______

3. Then it began to spin around very fast.

   ____________ ______ ____________ ______

4. The center came to be our sun.

   ____________ ______ ____________ ______

Check your answers on page 115.

Follow-Up

Work with a partner. How many of the nine planets can you name?

Do Continents Really Move?

What You Know People who read or write the Chinese language have to know at least 1,850 characters. Each character stands for a different idea or object. In the English language, we can read or write anything we want using only 26 letters.

How It Works

This lesson is about the sounds that are made by saying consonants. **Consonants** (KAHN-suh-nuhnts) are all the letters of the alphabet that are not vowels. The consonants are b, c, d, f, g, h, j, k, l, m, n, p, q, r, s, t, v, w, x, y, and z.

Each of the words below starts with a single consonant: *c*, *s*, and *r*. A vowel follows each consonant.

cold (c + o) **si**x (s + i) **ro**ck (r + o)

Some words start with two consonants in a row. For example:

from (fr + o) **sli**de (sl + i) **gri**p (gr + i)

Each consonant in these words has its own sound when you say the words out loud. We say the two sounds quickly, though, so they make almost one sound. Two consonants together like this are called **consonant blends**.

Here are some other words that start with consonant blends:

Word	Consonant Blend
blast	bl
crack	cr
scale	sc
twist	tw

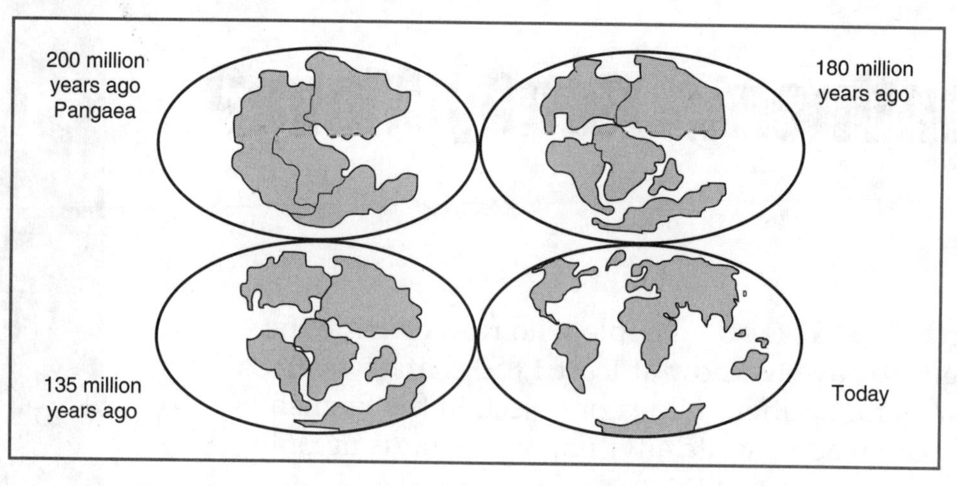

The continents have moved over the past 200 million years.

Try It

We live on the continent of North America. A **continent** (KAHN-tih-nihnt) is a large piece of land big enough to include many countries. Europe is a continent with many countries.

The world now has seven continents. Look at the picture above. Scientists think that a long, long time ago there was only one continent, called **Pangaea** (pan-GEE-uh). The picture shows how scientists think the continents have moved since that time. Circle the consonant blends at the beginning of these words:

small present stop

The consonant blends are **sm** in *small*, **pr** in *present*, and **st** in *stop*. Read these sentences. Circle the words that begin with consonant blends.

Pieces of land broke free from Pangaea.

Three words begin with consonant blends: **br**oke, **fr**ee, **fr**om.

These pieces slowly floated away from Pangaea.

Three of these words begin with consonant blends: **sl**owly, **fl**oated, and **fr**om. The *th* in *these* is not a consonant blend. The *t* and *h* do not have their own separate sound. When these two letters are together, they form a new, different sound.

Scientists call these pieces plates. Some are big and some are small.

Two of these words begin with consonant blends. They are **pl**ates and **sm**all.

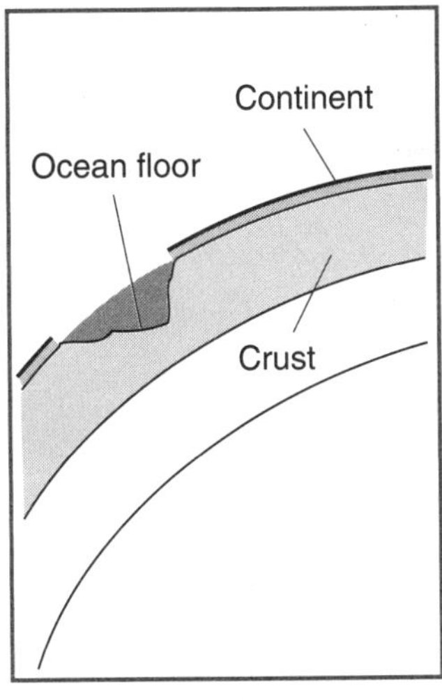

The outer part of the earth is called the **crust**.

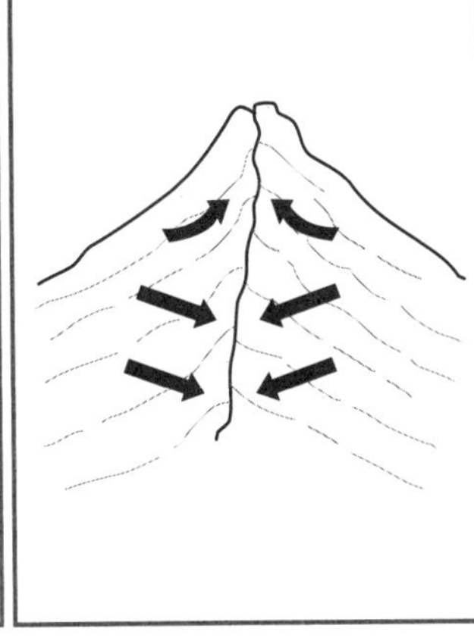

When two plates come together, a mountain can form.

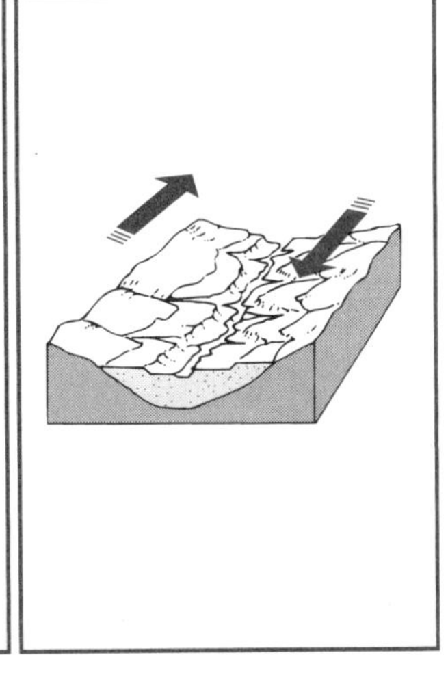

Earthquakes happen when two plates slide past each other.

Practice

Study the picture carefully. Then read the following sentences. Circle the consonant blends at the beginning of words.

1. The outer part of the earth is like its skin. This skin is called the earth's crust.
2. A plate is a broken piece of this crust.
3. Some plates are continents. Others are parts of the ocean floor.
4. The United States is on the American plate. It is slowly floating to the west.
5. Sometimes two plates crash into each other. This is how some mountains are formed.
6. Sometimes two plates slide by each other. This can cause an earthquake.

Check your answers on page 115.

Follow-Up

Can you point out the seven continents on a world map?

Lesson

3

How Is Rock Formed?

What You Know You know how to add new beginnings to a word to change its meaning. Think about tying your shoes. You can *tie*, *retie*, and *untie* them. The word beginnings *re* and *un* change the meaning of the word *tie*.

How It Works

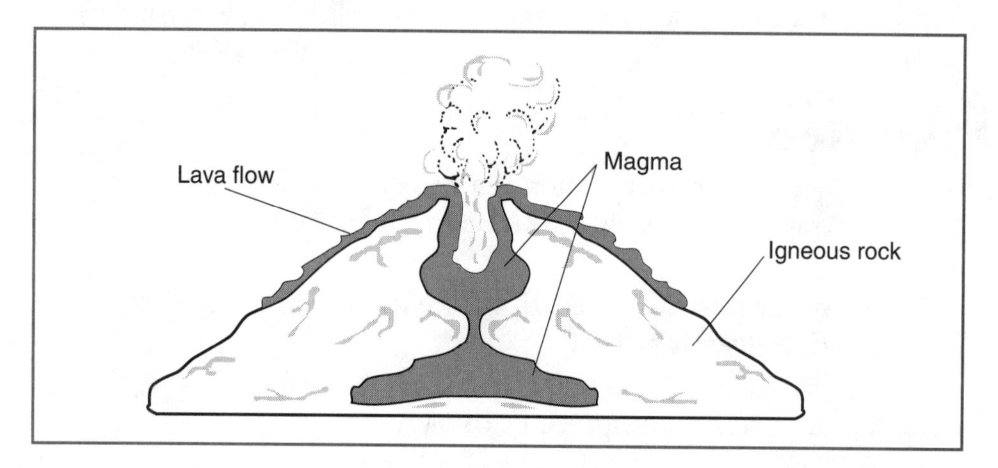

This is a drawing of a volcano.

The picture above shows where one kind of rock comes from. **Igneous** (IG-nee-uhs) rock comes from volcanoes. Pools of very hot melted rock are deep under the surface of the earth. This superheated rock is called **magma** (MAG-muh). The great heat forces the magma to rise. When the magma reaches the surface of the earth, it overflows.

Look at the underlined words above. They are *superheated* and *overflows*. The word beginnings on each word change its meaning. These word beginnings are called **prefixes** (PREE-fihks-ehz).

The prefix *super* means "very much" or "more than usual." So the word *superheated* means "very hot." Some jet planes are *superfast*. Can you guess what *superfast* means? It means "very fast."

The prefix *over* means "over the limit." There was too much magma to stay underground. So the magma *overflowed* — it flowed over its limits. This is similar to the way a toilet overflows. Sometimes people *overeat*. What does this mean? It means that they eat over their limit, or more than they should.

Try It

Now let's look at three more prefixes: *trans, in,* and *im.*

Trans means "across" or "through." A radio or television transmitter sends messages through space.

The prefixes *in* and *im* both mean "not." An *in*complete story is one that is not complete. An *im*movable rock is one that cannot be moved.

Put a check mark next to the words that complete the sentences correctly.

1. Magma comes from deep inside the earth. It is ________ through a tunnel to the surface.

   ______ imported

   ______ transported

   ______ superported

   The correct answer is *transported*. The magma is carried *through* a tunnel to the earth's surface.

2. Some volcanoes overflow or **erupt** (ee-RUPT) only once. Others erupt more than once. A volcano that erupts many times seems to have ____________ power.

   ______ supernatural

   ______ transnatural

   ______ innatural

   The correct answer is *supernatural*. The volcano seems to have great power.

3. Sometimes a volcano does not erupt for many years. Some volcanoes have not erupted for several hundred years. Such volcanoes are ____________.

   ______ overactive

   ______ inactive

   ______ transactive

   The correct answer is *inactive*. A volcano that has not erupted for a long time is *not* active.

This picture shows sedimentary rock.

Practice

Circle the prefix in each underlined word. Then put a check mark next to the correct meaning of the prefix.

1. Sedimentary (sehd-uh-MEHN-tuh-ree) rock is made from several different types of rocks. The rocks change a lot from the way they start out. Sometimes it is impossible to pick out the original rocks.

   ______ through

   ______ not

   ______ across

2. A very common type of sedimentary rock is sandstone. It is used in buildings all across the country. Special types of sandstone are sometimes transported from Vermont all the way to California.

   ______ very much

   ______ not

   ______ across

Check your answers on page 115.

Follow-Up

Do you know whether there are any volcanoes in the United States? How do you think you could find out?

What Kind of Cloud Is That?

What You Know When you see a movie with a friend, you may not agree about it. You say, "That was the best movie I've ever seen. It was the *greatest*!" Your friend says, "What? That was the *worst* movie of the year!" *Best* and *greatest* mean the same thing. *Best* and *worst* are opposites.

How It Works

Synonyms (SIHN-uh-nihmz) are words that mean the same thing as one another. **Antonyms** (AN-tuh-nihmz) are words that mean the opposite of one another.

These are cumulus clouds.

Cumulus (KYOOM-yuh-luhs) **clouds** usually appear low in the sky. They look flat on the bottom and rounded on top.

Read these sentences. Look at the underlined words.

1. Cumulus clouds are <u>large</u> and fluffy.
2. Cumulus clouds are <u>big</u> and fluffy.
3. Cumulus clouds aren't <u>small</u> and fluffy.

Sentence 2 uses the word *big* instead of the word *large*. *Big* is a synonym for *large*. It means the same thing as *large*.

Sentence 3 says that cumulus clouds aren't *small*. *Small* is an antonym for *big*. It means the opposite of *big*.

Try It

Read these three sentences. Look carefully at the underlined word in the first sentence.

1. Cumulus clouds often form on hot summer days.
2. These clouds frequently form on hot summer days.
3. These clouds rarely form on hot summer days.

Which word is a synonym for the word *often*?

____

Frequently is a synonym for *often. Frequently* means the same thing as *often.*

Which word is an antonym for *often*?

____

Rarely is an antonym for *often. Rarely* means the opposite of *often.*

Read these three sentences. Look carefully at the underlined word in the first sentence.

1. Cumulus clouds are usually a sign of fair weather.
2. Cumulus clouds are usually a sign of nice weather.
3. Cumulus clouds are not usually a sign of rainy weather.

Which word is a synonym for the word *fair*?

____

Nice is a synonym for *fair. Nice* weather is the same thing as *fair* weather.

Which word is an antonym for *fair*?

____

Rainy is an antonym for *fair. Rainy* means the opposite of *fair.*

These are cirrus clouds.

Practice

Answer the questions that follow each group of sentences.

1. Cirrus (SIHR-uhs) clouds are made of tiny pieces of ice. These pieces are so small you need a microscope (MEYE-kroh-skohp) to see them. Some cirrus clouds are very large.

   a. Which word is a synonym for *tiny*? ____________

   b. Which word is an antonym for *tiny*? ____________

2. Cirrus clouds often appear when the weather is cool and dry. They don't usually appear in warm weather. You can often see cirrus clouds on chilly fall evenings.

   a. Which word is a synonym for *cool*? ____________

   b. Which word is an antonym for *cool*? ____________

Check your answers on page 115.

Follow-Up

Look out a window or go outdoors. Can you locate any cumulus or cirrus clouds?

Lesson

5

Why Does the Wind Blow?

What You Know You would never say, "Turn on the *land*" when you meant, "Turn on the *lamp*." Or "I want a glass of *mint*" when you meant, "I want a glass of *milk*." You automatically choose the correct pair of letters to finish words like these.

How It Works

Lessons 1 and 2 were about the sounds of the vowels and consonants. Look back at Lesson 2. It gives examples of words that start with consonant blends.

The two separate sounds in a consonant blend are said quickly, one after the other. They almost form a single sound. Some examples of consonant blends at the beginnings of words are:

from **sl**ide **gr**ip

This lesson is about words that *end* with a consonant blend. The examples from above are:

la**nd** la**mp** mi**nt** mi**lk**

Say these words out loud. You pronounce the last two letters together. Each consonant has its own sound. When you say them together quickly, the two sounds almost form a single sound.

Here are some other words that end in consonant blends:

Word	Consonant Blend
end	nd
melt	lt
desk	sk
rent	nt

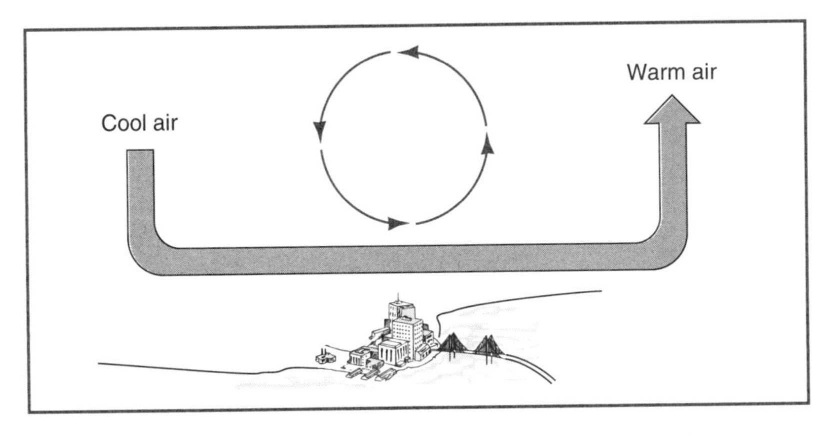

This drawing shows how a convection current works.

Try It

Circle the consonant blends at the end of these words:

list band lift held

The consonant blends are **st** in *list*, **nd** in *band*, **ft** in *lift*, and **ld** in *held*.

Circle the consonant blends at the end of the words below.

1. You have felt the heat of the sun on your face.
2. The sun also heats the land and water on earth.

Four words end with consonant blends. They are fe**lt**, hea**ts**, la**nd**, and a**nd**. Each letter of the consonant blend keeps its own sound.

The word *earth* ends in two consonants, *t* and *h*. However, these letters do not form a consonant blend. When you say *t* and *h* together, neither letter keeps its own sound. They form a new sound.

Now read these sentences. Circle the consonant blends at the ends of words.

1. The air can also hold some heat from the sun.
2. This hot air rises away from the earth. Cool air moves into the space left behind.
3. This movement of air is called a **convection** (kuhn-VEHK-shuhn) current.

Which words end with consonant blends? They are ho**ld**, le**ft**, behi**nd**, moveme**nt**, and curre**nt**.

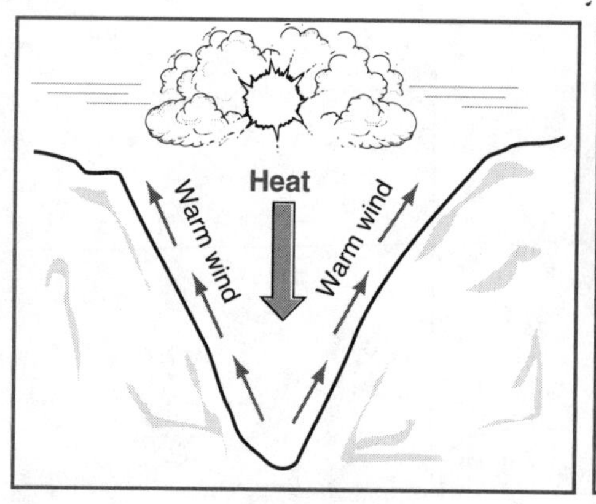

Warm winds blow upward on the mountains on a sunny day.

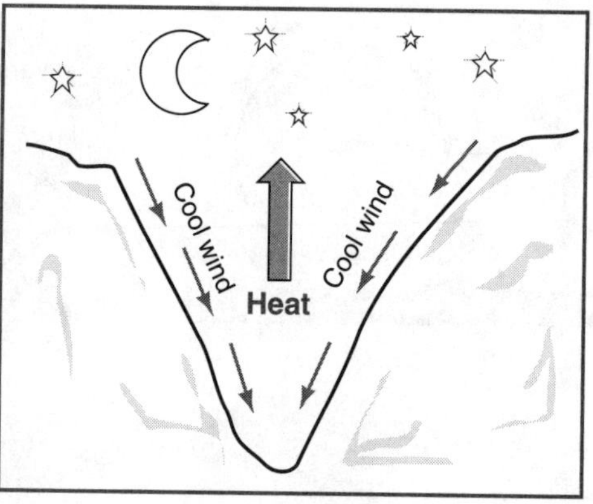

Cool winds blow downward on the mountains at night.

Practice

Copy the underlined words into the answer blanks below. Circle the consonant blends at the ends of these words.

1. A mountain can have an important effect on our weather.

   ____________ ____________

2. A warm air current blows upward on a mountain on a sunny day.

   ____________ ____________ ____________

3. A cold mountain wind blows downward on a mountain at night.

   ____________ ____________ ____________

4. These winds act like rivers of air. They change the weather from hour to hour.

   ____

Check your answers on page 115.

Follow-Up

Make a list of different kinds of winds you have felt (cold, warm, strong, gentle).

How Tall Are Mountains?

Lesson

6

What You Know You can understand some kinds of information more quickly in picture form than in words. For example, ads may compare three kinds of flashlight batteries by using pictures that show lines in them. The longest line is shown next to the longest-lasting battery. The shorter lines are next to the names of other batteries. We know right away that we should buy the battery with the longest line next to it.

How It Works

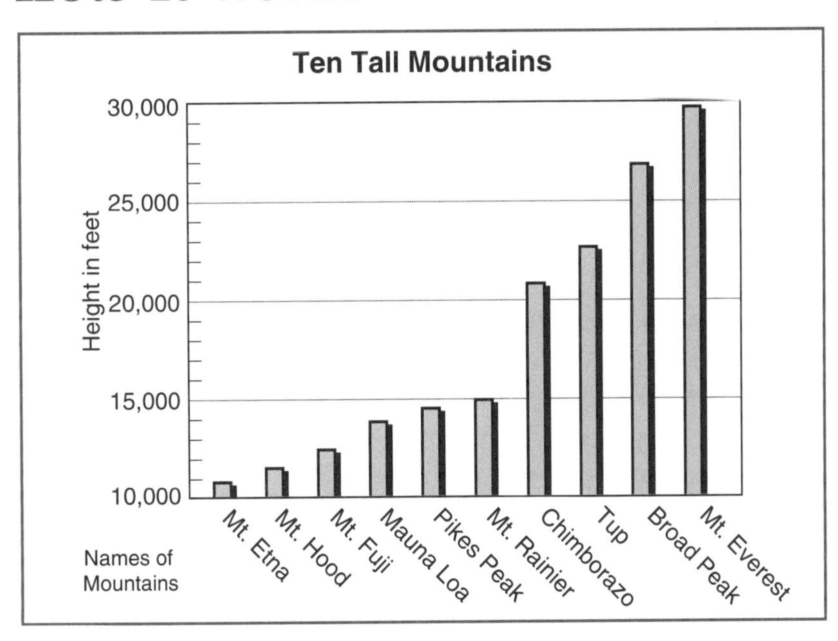

This bar graph shows mountain heights.

The chart above is called a **bar graph**. It uses thick lines called **bars** to compare different things.

First look at the title of the graph. It tells you what the graph shows. The title is "Ten Tall Mountains." This tells you that the graph gives information about ten tall mountains.

Next look at the words at the very bottom. There is a small title, "Names of Mountains." This tells you that the words above this title (shown sideways) are the names of different mountains.

Now look at the left side of the graph. At the top is a small title that says "Height in feet." This means that the numbers below the title tell how many feet high each mountain is. The numbers start at the bottom with 10,000 and go up to 30,000 at the top. This means that the heights start at 10,000 feet and increase by jumps of 5,000 feet.

Lines go across the graph. They appear after every 5,000 feet. The small lines on the left side of the graph are used for every 1,000 feet.

These lines help you find numbers on the graph. For example, to find 17,000 feet, you can count two small lines up from the 15,000-foot mark.

Try It

Put a check mark next to the correct answer below.

1. How tall is Mount Rainier?

______ A. about 15,000 feet

______ B. about 20,000 feet

______ C. about 25,000 feet

The correct answer is A. Find Mount Rainier at the bottom of the graph. Then follow the bar above Mount Rainier from the bottom up to where it ends. Put your finger on the nearest line going across. Follow that line back to the numbers at the left. That line is the 15,000-foot line.

2. Which mountain is just over 12,000 feet tall?

______ A. Mount Etna

______ B. Mount Fuji

______ C. Tup

The correct answer is B. First find the 12,000-foot line at the left of the graph. It is the second line up from the bottom. Follow this line with your eye across the chart. Look for a bar that ends just above this line. The third bar, Mount Fuji, ends just above the 12,000-foot line.

Practice

Put a check mark next to the correct answer to each of the questions.

1. What is the shortest mountain shown on the chart?

_____ A. Mount Hood

_____ B. Mount Etna

_____ C. Broad Peak

2. How tall is Mount Hood?

_____ A. About 1,000 feet

_____ B. About 11,000 feet

_____ C. About 21,000 feet

3. Which two of the mountains are about 14,000 feet tall?

_____ A. Pikes Peak and Mauna Loa

_____ B. Chimborazo and Tup

_____ C. Pikes Peak and Chimborazo

4. Broad Peak is ____________ Mount Rainier.

_____ A. taller than

_____ B. shorter than

_____ C. about the same height as

Check your answers on page 115.

Follow-Up

Have you ever been to a mountain? If so, do you know how tall it was? How do you think you can find out?

Unit 2

Biology

Plant and Animal Cells

What You Know After you have watched a soap opera several times, you can usually guess what each person will do. The same is true of saying new words. When a new word looks like one you have seen before, the two words often sound the same.

How It Works

In Lesson 1 you learned about long and short vowels. All the vowels—*a*, *e*, *i*, *o*, and *u*—have both short and long sounds.

Think of the words *rat* and *rate*. The *a* in *rat* is a short vowel. It doesn't sound like the *a* in the alphabet. The *a* in *rate* is a long vowel. It sounds like the letter *a* in the alphabet.

Now look at these pairs of words. The first vowel in each pair is the same. However, the vowel *sounds* in each pair are different.

take	sat	slow	top
be	red	huge	but
like	lick		

Try It

Look at the underlined words in items 1 and 2 below. In some of these words, the first vowel is a short vowel. It doesn't sound like the letter in the alphabet. Write only these words in the answer blanks.

1. All living things are <u>made</u> <u>of</u> <u>cells</u>.

____________ ____________

The *o* in *of* and the *e* in *cells* are short vowels. The *a* in *made* is a long vowel.

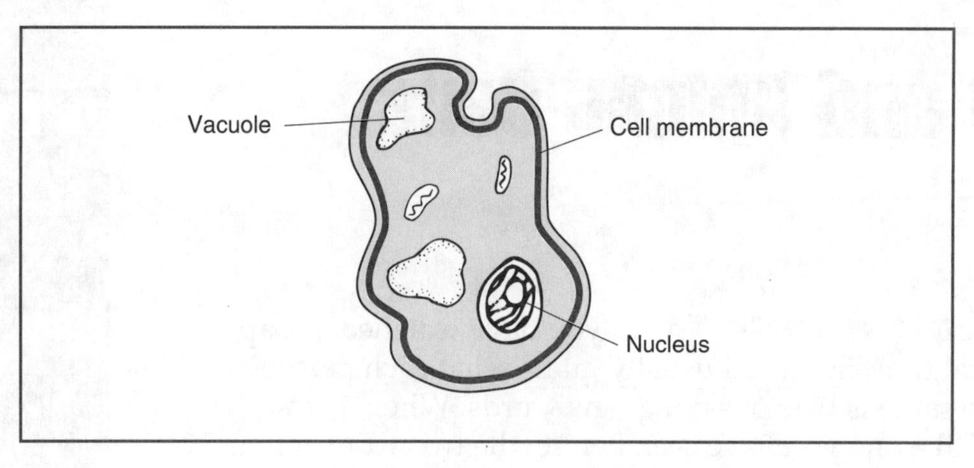

This is a drawing of a cell.

2. The skin of a cell, called the cell membrane (MEHM-brayn), holds the cell together.

____

The *i* in *skin* is a short vowel. The *o* in *holds* is a long vowel.

Now look at the underlined words in items 3, 4, and 5. In some of these words, the first vowels are long vowels. Write only these words in the answer blanks.

3. At the same time, the membrane lets food pass into the cell.

__________ __________

The *a* in *same* and the *i* in *time* are long vowels. The *e* in *lets* and the *a* in *pass* are short vowels.

4. The membrane also allows waste to be passed out of the cell.

__________ __________

The *a* in *waste* and the *e* in *be* are long vowels. The *o* in *of* and the *e* in *cell* are short vowels.

5. The vacuoles (VAK-yoo-ohlz) are big open spaces where food and waste can be stored.

__________ __________

The *a* in *spaces* and the *e* in *be* are long vowels. The *i* in *big* and the *a* in *and* are short vowels.

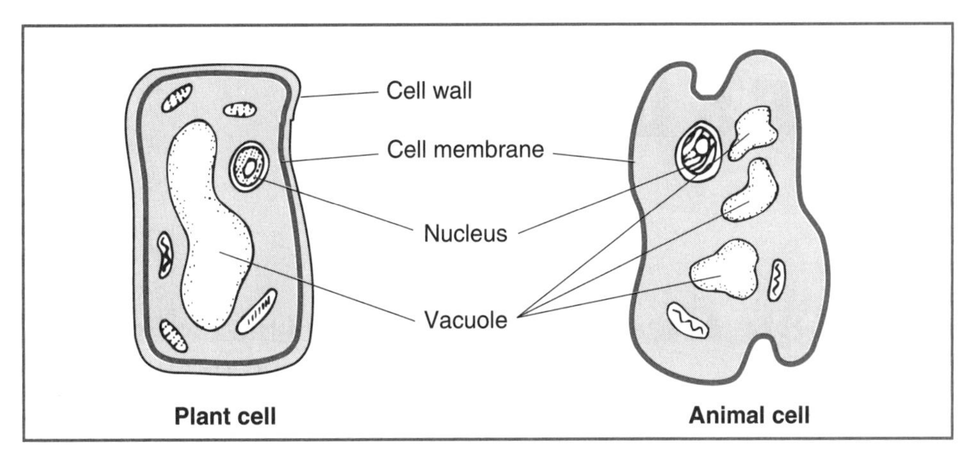

This drawing shows a plant cell and an animal cell.

Practice

Copy each underlined word on one of the long answer blanks. Look at the first vowel in the word. Write *long* on the short answer blank if the word has a long vowel sound. Write *short* on the short answer blank if the word has a short vowel sound.

1. Both plant and animal cells are building blocks of life.

   ____________ ______ ____________ ______

2. Plant cells can use sunlight to make food.

   ____________ ______ ____________ ______

3. Animal cells cannot make food like plant cells can.

   ____________ ______ ____________ ______

4. Plant cells usually have one or two very big vacuoles.

   ____________ ______ ____________ ______

5. Animal cells usually have several little vacuoles.

   ____________ ______

Check your answers on page 115.

Follow-Up

Reread the information about cells in the Try It and Practice sections of this lesson. Can you find something that people and cells have in common?

Lesson

8

Darwin's Discovery

What You Know Most members of the Pro Football Hall of Fame had *long* playing careers. Tony Dorsett put in 12 seasons for the Cowboys and Broncos. Terry Bradshaw played even *longer*—14 seasons with the Steelers. Johnny Unitas had the *longest* career of all—18 seasons with the Colts! The word endings *er* and *est* change the meaning of the word *long*.

How It Works

Once some giraffes had short necks. They couldn't reach as many leaves as giraffes with long necks. Today all giraffes have long necks.

Here are three common word endings: *ist*, *er*, and *est*. Word endings like these are called **suffixes** (SUHF-iks-ehz). They change the meaning of the word they are added to.

The suffix *ist* means "a person who knows a lot about something."

> Darwin was a scient**ist** who sailed around the world in 1831. (A *scientist* is someone who knows a lot about science.)

The suffix *er* means *more*. It is used when you compare two different things. To *compare* is to look at two things to see how they are alike and how they are different. A word with *er* at the end is more than the word it is compared with.

> He thought about how animals had changed over the years. He decided that some giraffes once had short**er** necks than others. (This means that the necks of some giraffes were "more short" than others—the giraffes with long necks.)

You just learned that the suffix *er* means *more*. The suffix *est* means *most*. It is used only when you are comparing more than two things.

The giraffes with the long**est** necks could reach leaves that no other giraffes could reach. Darwin believed that short-necked giraffes died out. Those with the long**est** necks lived on. (This means that giraffes that had necks that were the "most long" were the ones that lived.)

If a word ends in *e*, do not add another *e* to the word. Just add an *r* for the suffix *er* and *st* for the suffix *est*.

Try It

Read the sentences below. Look at the underlined words. Pick the suffix that best fits the meaning. Then write the suffix in the space.

Suffixes: er est ist

1. Darwin was a well-known natural______. He knew a lot about nature.

Darwin was a natural*ist*. The suffix *ist* means "a person who knows a lot about something."

2. Many people have studied Darwin's ideas. One person found that dark-colored rabbits are safe_____ in the forest than light-colored rabbits. The dark-colored rabbits blend in with the background. The light-colored rabbits do not. They can be seen more easily by animals that hunt rabbits.

The *er* in *safer* means "more safe."

3. Animals that could run the fast_____ could get away from animals that were hunting them.

The *fastest* animals were the ones that got away. They were the "most fast."

This is a picture of different colored rabbits.

Practice

Read the sentences below. Look at the underlined words. Pick the suffix that best fits the meaning. Write the suffix in the space.

Suffixes: er est ist

1. Dark rabbits are hard_____ to see in the forest than light ones. Therefore, animals that hunt dark rabbits can't find them.

2. Look at the picture above. The art_____ made one rabbit completely white.

3. Brown rabbits are safe_____ in the forest than white ones because it's hard to hunt what you can't see.

4. Horses have also changed over the years. In the past, they had four toes. Now they have one big hoof. They can run fast_____ now than they could before.

5. The horses with the longest legs can run the fast_____ of all.

Check your answers on page 116.

Follow-Up

Have you ever seen a polar bear at the zoo? Why do you think it's white? Hints: Polar bears live where there is lots of snow and ice. Polar bears are hunters.

Keeping Healthy

What You Know When you watch television, you sometimes hear a word you have never heard before. By fitting the word with other words and sentences around it, you usually can guess what the new word means. You can learn new words the same way when you read.

How It Works

Good **nutrition** (noo-TRIH-shuhn) is the key to good health. We all need to watch how much we eat. Eating the right kinds of food is just as important.

Do you know what *nutrition* means? Here's how to figure it out:

1. Look at the word at the end of the first sentence above. It is *health*. The word *health* tells you that *nutrition* has something to do with health.

2. Look at the second sentence. The words *how much we eat* tell you that *nutrition* has something to do with the amount of food we eat.

3. Look at the last sentence. The words *the right kinds of food* tell you that *nutrition* has something to do with eating one type of food instead of another.

So *nutrition* is the kind and the amount of food a person eats.

Try It

Read the following sentences. Then guess what **vitality** (veye-TAL-uh-tee) means. Use the words around it for clues, as you did with the word *nutrition*.

First look at the words near the word you don't know. Second look at the sentences near the unknown word. Then make your guess.

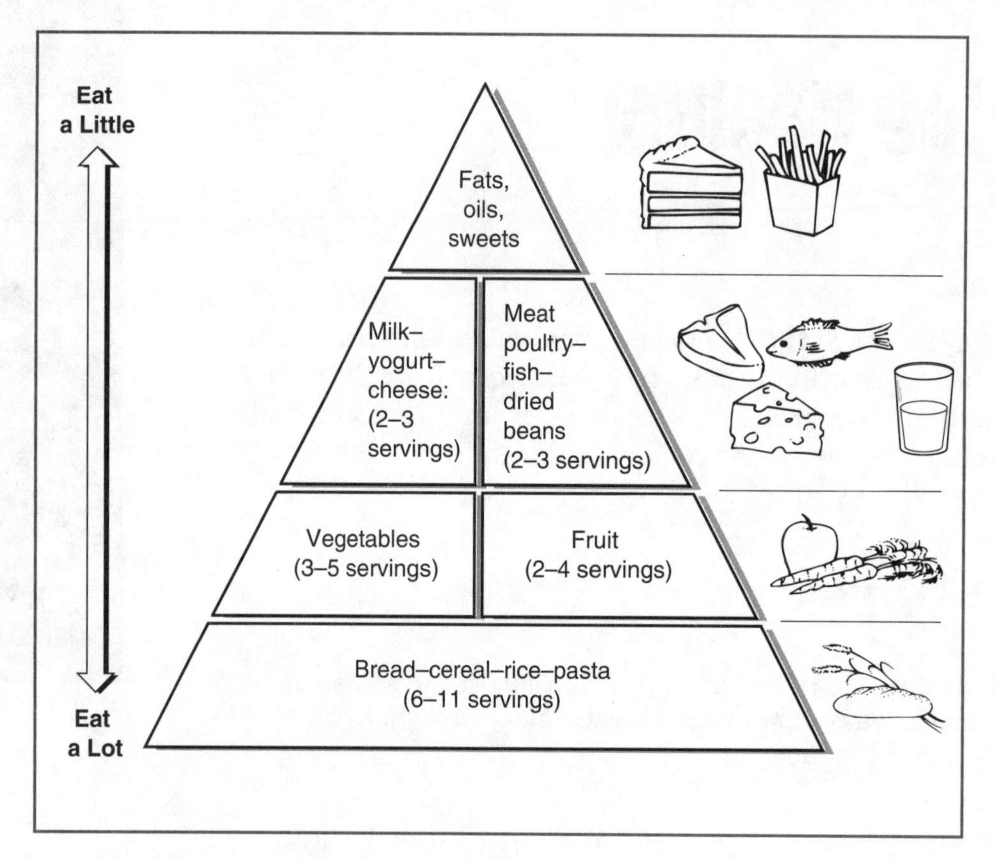

The food guide pyramid shows different kinds of food. To eat a healthful diet, eat more of the foods shown toward the bottom. Eat less of the foods shown toward the top.

A good diet is one way to increase your *vitality*. When you eat healthy foods, you have more energy. You don't get tired quickly.

What does *vitality* mean? Write the meaning on the answer blanks.

Vitality means having energy and not getting tired quickly.

1. Look at the other words in the first sentence. It shows that a "good diet" *increases vitality*.

2. Look at the second sentence. The sentence says that eating healthful foods—which is the same thing as having a "good diet"—gives you more energy. So part of *vitality* is having more energy.

3. Look at the third sentence. The words *don't get tired quickly* show that *vitality* has to do with not getting tired fast.

So *vitality* means having a lot of energy and not getting tired quickly.

Practice

Read the sentences below. Then figure out what the underlined word in each sentence means. Write the meaning on the answer blanks below each sentence.

1. Good nutrition helps in building resistance to disease. It helps you fight off diseases before they start. A healthy diet helps you build up your strength so that you get sick less often.

   ____

   ____

2. Doing aerobics (uh-ROH-bihks) is another way to keep healthy. Swimming, walking, playing sports, bicycle riding, and even dancing make your heart stronger and help you breathe better.

   ____

   ____

Check your answers on page 116.

Follow-Up

Write down everything you eat for one 24-hour period. Which foods are healthful? Which are not?

Lesson 10

The Circulatory System

What You Know

When your favorite team is playing in an important game, there are certain kinds of information you want to know. You may look in the sports section of a newspaper to find this information. For important games, your newspaper probably will have lots to say and all kinds of information.

You may read all the information and pick out the details you want. The details may include who will be starting, what time the game will be on TV, and whether your favorite player is well enough to play.

How It Works

When you read to learn something, rather than just for fun, you will find it helpful to ask yourself questions as you read. The questions may be, "Who did that?" or "When did that happen?" Then, as you read, you can look for the answers.

First you need to think about what you need to know. Next ask yourself a question that starts with *who, what, where, when, why*, or *how*.

You can find the answer to your question in what you are reading. The answer can often be found in a single word or just a few words. The answer may be a place, a date, a time, a number, a person, or an explanation.

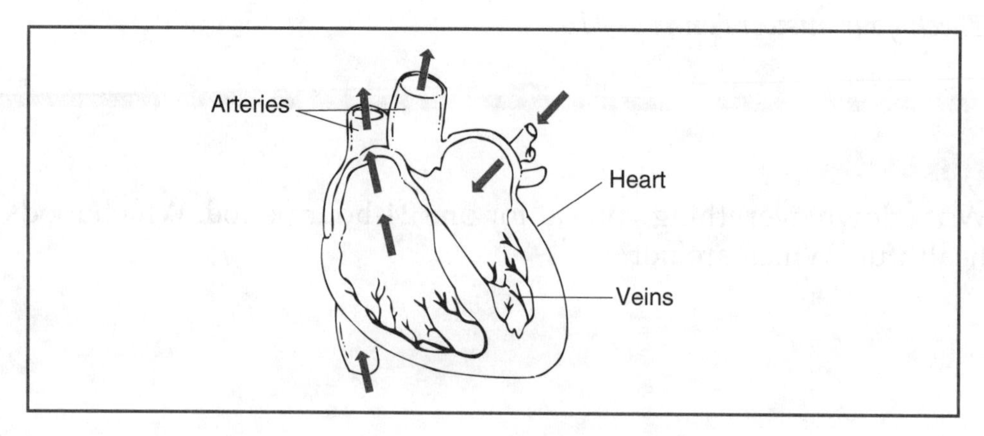

This shows how the human heart circulates (SER-kyoo-layts) blood.

Try It

Read these sentences. Then answer the questions.

William Harvey was the first person to discover how blood **circulates**, or moves, throughout the body. Harvey lived in England from 1578 to 1657. He studied living animals to find out how the heart works. He first described how the heart pumps blood to all parts of the body in 1616.

1. Who discovered how blood circulates throughout the body?

____

____

William Harvey discovered how blood circulates. (The question starts with *who*. The answer is a person's name.)

2. Where did Harvey live?

____

____

Harvey lived in England. (The question starts with *where*, and the answer is the name of a place.)

3. When did Harvey first describe how blood circulates?

____

____

He first described circulation (ser-kyoo-LAY-shuhn) of blood in the year 1616. (The question starts with *when*. The answer is a date.)

4. How did Harvey find out how the heart works?

____

____

He studied living animals. (The question starts with *how*. The answer is an explanation.)

Practice

Read these sentences. Then answer the questions.

A baby's heart begins to beat several weeks after it is conceived. The blood enters the right side of the heart. This side then pumps the blood into the left side. The left side of the heart then pumps the blood out to all parts of the body.

1. When does a baby's heart begin to beat?

____

____

2. Where does the blood enter the heart?

____

____

3. What does the right side of the heart do?

____

____

4. How does the blood reach the rest of the body?

____

____

Check your answers on page 116.

Follow-Up

Have someone show you how to take your own pulse rate. Your pulse rate shows how many times a minute your heart beats.

Some Healthy Lessons

Lesson 11

What You Know Have you ever made a cake from a mix by following the directions on the outside of the box? Have you gotten your VCR to work so that it would record your favorite program when you were not home? You did these things by studying the directions and following them carefully.

How It Works

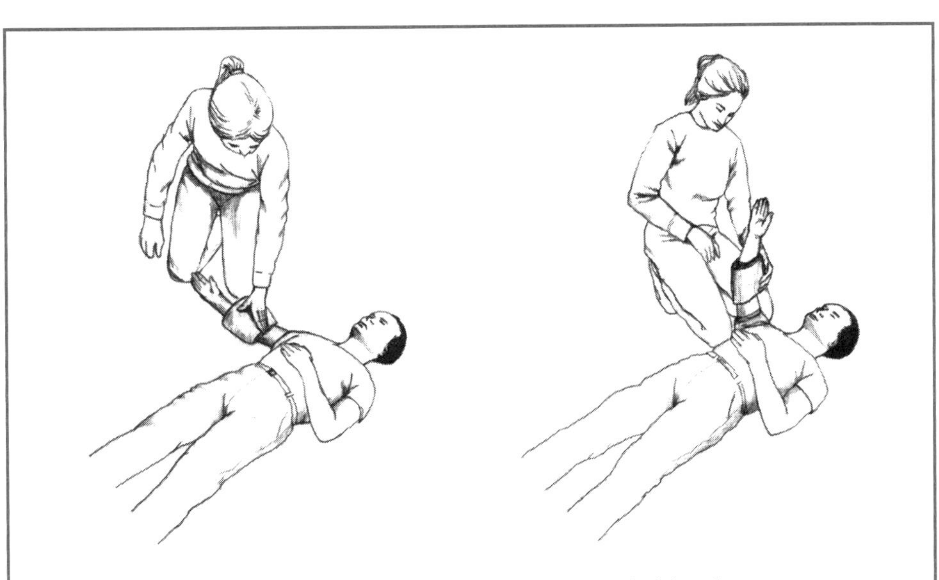

This shows how to help someone who is bleeding.

1. Help the person who is bleeding sit or lie down.
2. Press a clean cloth to, or wrap a cloth around, the bleeding area.
3. If possible, raise the area that is bleeding above the rest of the person's body.
4. Put a fresh cloth over the first one when it becomes soaked.
5. Call for help as soon as possible.

Do you know what to do if someone you are with is hurt and needs help? Look at the pictures and steps to follow shown above. To help someone who is hurt or to follow any other directions, you must do certain things.

1. Before you begin, read the directions from beginning to end. Get a general idea of what you are going to do. (There are five basic steps in helping someone who is hurt and bleeding.)

2. Figure out words you don't understand. (In step 4, what does *soaked* mean? Here it means "filled with blood.")

3. Read the steps again and picture doing each one in your mind. (For example, figure out how you would wrap the person's arm as shown in step 2.)

Try It

Read these directions on how to lift a heavy box. Then answer the questions.

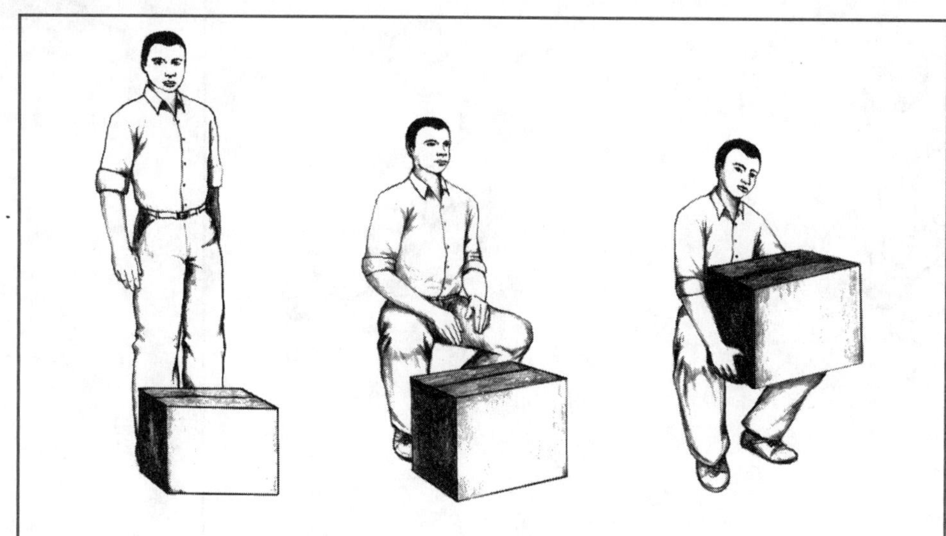

This shows how to lift a heavy box.

1. Stand close to the box.
2. Place your feet wide apart.
3. Place one foot a little ahead of the other.
4. Bend your knees. Do not bend at the waist.
5. Use your leg muscles to lift the box.

1. What are the instructions about? Put a check mark next to the right answer.

   ______ A. How to help an injured person

   ______ B. How to save time

   ______ C. How to protect yourself

The correct answer is C. Following the instructions does not make the job go faster, and no one has been injured. These instructions are meant to keep you from getting hurt when you lift something heavy.

2. In step 2, what does *wide apart* mean?

______ A. About 2 to 4 inches apart

______ B. About 4 to 6 inches apart

______ C. About 1 to 2 feet apart

The correct answer is C. Most people would find their feet wide apart if they were 1 to 2 feet from each other. When your feet are 2 to 6 inches away from each other, they are not very wide apart. Also look to see how much 1 to 2 feet are.

Practice

Now answer these questions about lifting a heavy box.

1. In step 3, what does *a little ahead of* mean?

______ A. Place one foot a bit in front of the other.

______ B. Place one foot on top of the other.

______ C. Place both feet 3 inches from the box.

2. What do you do in step 5?

______ A. Lift the box by straightening your back.

______ B. Lift the box by straightening your legs.

______ C. Lift the box by straightening your arms.

Check your answers on page 116.

Follow-Up

Think of something you know how to do well. Write down each step you must take to do this. Then let a friend or classmate follow your instructions.

Lesson

12

Animals at Risk

What You Know You have probably seen several different types of maps. One kind of map shows the names of cities and states. Another type shows streets and highways. Some maps use pictures to give information. For example, a vacation-planning map might have a picture of Niagara Falls in New York State and the Grand Canyon in Arizona.

How It Works

This map shows at-risk animals in the United States.

People have taken a lot of land away from the animals that used to live on the land. Some types of animals that may soon die out are shown on this map.

This map shows two kinds of information:

1. It shows the United States (except for Alaska and Hawaii).

2. It shows pictures of several animals that are dying out. A line connects each animal to a state. For example, the map shows that some of the remaining Key Deer live in Florida.

Try It

1. According to this map, which state has the most types of at-risk animals?

   ______ A. New York

   ______ B. Texas

   ______ C. Florida

   The correct answer is C. There are three types of at-risk animals in Florida: the Key Deer, the Atlantic Salt Marsh Snake, and the American Crocodile (KRAHK-uh-deyel). None of the other states on this map has more than one.

2. Where is the Masked Bobwhite found?

   ______ A. Arizona

   ______ B. Alabama

   ______ C. California

   The correct answer is A. The Masked Bobwhite is found in Arizona.

3. Which at-risk animal lives in Utah?

   ______ A. the Prairie Dog (PREHR-ee DAWG)

   ______ B. the Jaguarundi (jag-wuh-RUHN-dee)

   ______ C. the American Crocodile

   The correct answer is A. The Prairie Dog lives in Utah.

Practice

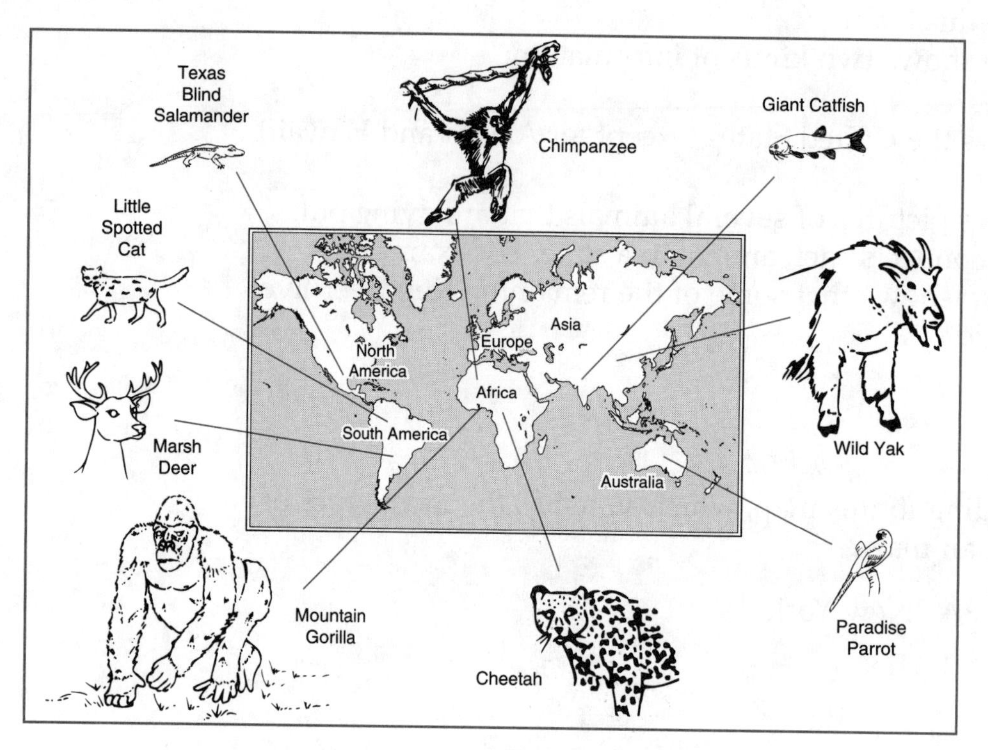

This map shows some of the at-risk animals in the world.

Use the map above to answer these questions. Place a check mark next to the correct answer.

1. Which continent on this map has the most types of at-risk animals?

_______ A. North America

_______ B. Asia

_______ C. Africa

2. Where do Giant Catfish live?

_______ A. North America

_______ B. Asia

_______ C. Africa

3. Which at-risk animal lives in Africa?

_______ A. the Cheetah (CHEE-tuh)

_______ B. the Marsh Deer

_______ C. the Wild Yak (YAK)

4. On which continent does the Paradise Parrot live?

_______ A. Africa

_______ B. Asia

_______ C. Australia

Check your answers on page 116.

Follow-Up

Many kinds of animals have died out since life began on Earth. Do you know of any? (Hint: A few years ago, a very popular movie had these animals in it. These animals are very big.)

Unit 3

Physics

Very Special Sounds

What You Know When you buy a frozen dinner in the supermarket, you probably don't take the time to read every word on the package. However, you do pick out the details that are important to you. These might include exactly what foods are included, whether they are fattening or not, how much food is included, and how long it takes to cook.

How It Works

In Lesson 10, when you looked for the important facts in paragraphs, you learned to follow two steps:

1. Decide what you need to know.
2. Ask yourself questions that start with *who, what, where, when, why,* or *how.*

The answer is sometimes in a single word or just a few words. It may be a place, a date, a time, a fact, a person, or an explanation.

Try It

The following paragraph is about sound. Read the paragraph and answer the questions that follow. Look back at the hints in the How It Works section of this lesson as you answer the questions. Write your answers on the answer blanks.

> Doctors can use sound to "see" an unborn baby. They do this with an **ultrasound** (UHL-truh-SOWND) **machine**. They place part of the machine on the woman's stomach. The machine sends out sound waves. These waves pass through the woman's body and bounce off everything inside her. The machine then takes pictures of the sound waves that bounce back. These "sound pictures" show the baby inside the woman. This is how doctors can safely examine a baby before it's born.

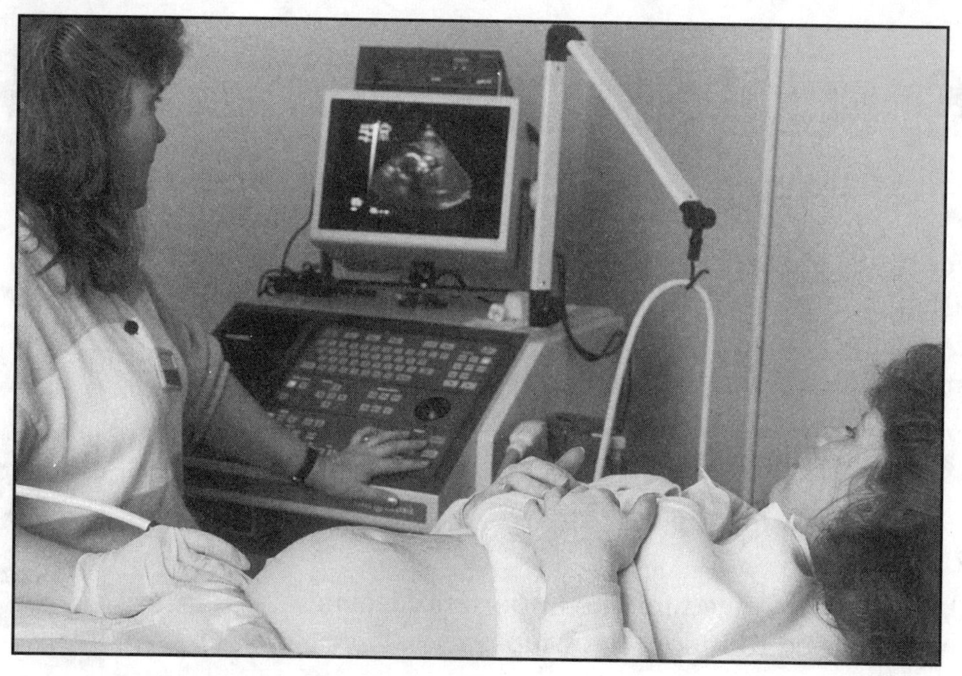

You can see a baby before it's born by using an ultrasound machine.

1. Who uses ultrasound machines? ____________________

____

Doctors use ultrasound machines. (The question starts with *who*. The answer is a group of people.)

2. Where does a doctor place the ultrasound machine?

____

The doctor places the machine on the woman's stomach. (The question starts with *where*. The answer is a place.)

3. When do doctors use ultrasound to examine a baby?

____

They use ultrasound to examine a baby before it's born. (The question starts with *when*. The answer is a time.)

4. What does the ultrasound machine take pictures of?

____

It takes pictures of sound waves that bounce off things inside the woman's body. (The question starts with *what*. The answer is a fact.)

Practice

Edison's recording machine looked like this.

Read this paragraph. Then answer the questions.

> Thomas Edison was the first person to record sound. The first sound recording was made in 1877. Edison spoke into a large cone that had a needle at the small end. The sound waves made by his voice made the needle move. The needle made marks on a roll of wax that was turning around. After he finished speaking, he placed the needle at the beginning of the roll of wax. Then he turned the crank, and he heard his voice.

1. Who made the first sound recording? ____________

2. When was the first sound recording made? ____________

3. What did Edison speak into? ____________

Check your answers on page 116.

Follow-Up

On the first sound recording machine, Edison spoke into a cone. Today, people who make recordings speak (or sing) into a microphone. In Edison's time, the sound was recorded on a wax roll. Can you guess what we use today instead of the wax roll?

Lesson

14

How Light Works

What You Know You may love hot, sunny weather. You walk out the door into the heat and say, "What a beautiful day!" Your friend may hate hot weather. He feels the heat on his face and says, "What awful weather!" You and your friend have different opinions about the weather. But it's a fact that the temperature is 90 degrees.

How It Works

You can see that a lamp is on the table. This is a *fact.* Do you like the lamp? Whether you do or not is your *opinion.*

It is important to know the difference between facts and opinions. You can count, measure, or see facts.

- There is only one lamp in the room.
- There is a 100-watt bulb in the lamp on the table.
- The lamp is turned on.

Facts can be proven to be real or true. They do not change.

- The lamp needs electricity to work.
- The bulb in the lamp is not red.

Opinions cannot be proven to be true. They are based on values, or a point of view. One way you can tell that something is a value or point of view is by looking for key words, such as *very, too, good, bad, nice,* and *awful.*

- That lamp is too small.
- It doesn't give good light.

Opinions may also be based on feelings or beliefs. These opinions often start with *I like* or *don't like, I think, I believe,* or ______ *should.*

- I don't like that lamp.
- I think we should get a new one.

There can be many different opinions about the same thing.

- Ellie thinks that the lamp is really pretty.
- Barry doesn't like the lamp at all.

Try It

Write F in the answer blank if the sentence is a fact. Write O if the sentence is an opinion.

___ **1.** The lamp on the table is made of metal.

Fact. You can see that the lamp is made of metal.

___ **2.** The lamp is very big.

Opinion. The statement is based on one person's view of what is big. Notice the word *very.*

___ **3.** That lamp isn't plugged in.

Fact. You can see that the plug is not in an outlet.

___ **4.** I think we need a second lamp in this room.

Opinion. This statement shows one person's beliefs. Notice the words *I think.*

Practice

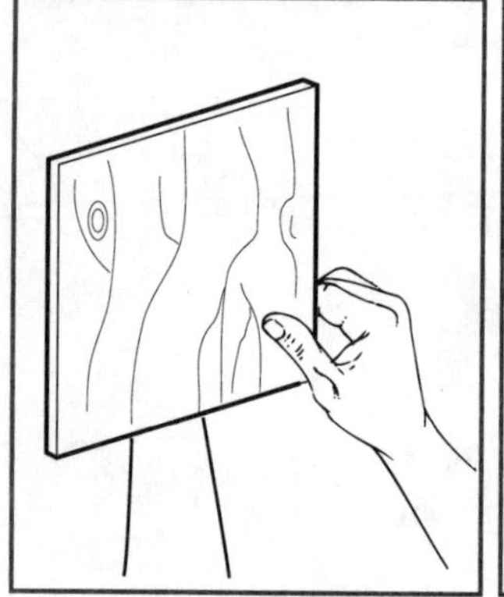

Opaque. The square is made of wood. Nothing shows through wood.

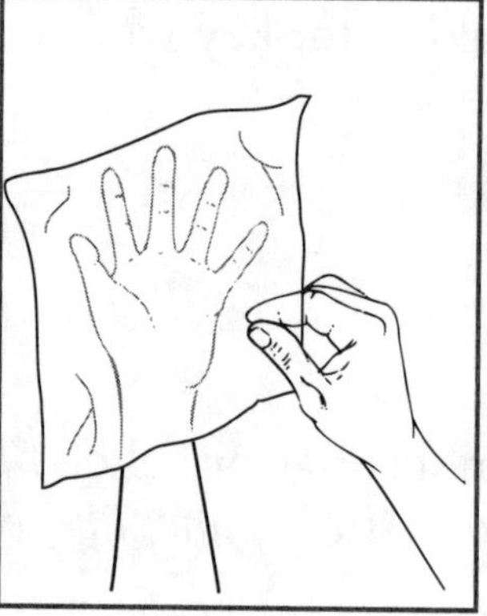

Translucent. The square is wax paper. The outline of the left hand can be seen through the paper.

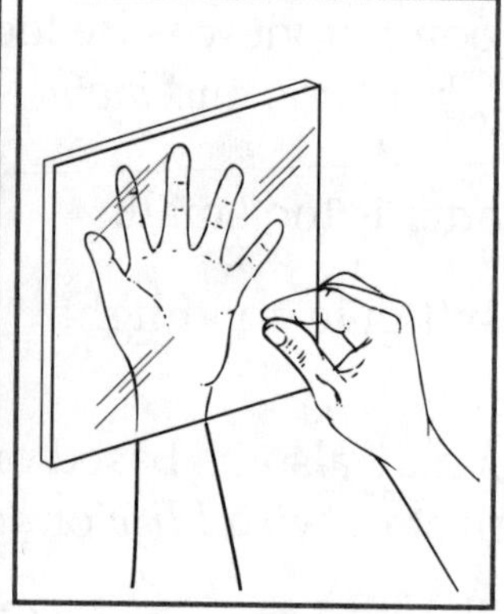

Transparent. The square is glass. The details of the left hand can be seen through the glass.

Light passes through glass easily because glass is **transparent** (tranz-PAHR-uhnt). It does not stop light. Other materials, such as notebook paper, stop some but not all light. Materials that allow some light to pass through are **translucent** (tranz-LOO-suhnt). A brick is **opaque** (oh-PAYK). It lets no light through.

Look at the following sentences. Write F if the sentence is a fact. Write O if the sentence is an opinion.

___ **1.** Wood isn't transparent.

___ **2.** Translucent window shades look nice.

___ **3.** Translucent objects let some light pass through.

___ **4.** I don't like the opaque windows on Bill's car.

___ **5.** Transparent raincoats are very popular.

Check your answers on page 116.

Follow-Up

Look around you. Can you see one thing that is transparent? One thing that is translucent? One thing that is opaque?

Measuring Up

Lesson

15

What You Know

You decide to give a birthday party for a friend. To get ready, there are many things you must do. For instance, you may buy a gift, invite guests, plan what food you will serve, and make a cake. These four things form a list. The main idea of the list is "Getting ready for a birthday party."

How It Works

Most lists are about one thing. For example, *buy a rug*, *fix a lamp*, and *make a coffee table* are three things you might do to fix up your home. The items on this list all have one thing in common. What they have in common is the **main idea** of the list.

The list of words below has one main idea. The main idea is *length*. What the words have in common is that they are all measures of length.

inches

meters

length

centimeters

The next list also has a main idea. The main idea is *speed*. The other items are different ways of measuring speed. The speed of a car can be measured in miles per hour or (in Canada and Europe) kilometers per hour. The speed of rockets is sometimes measured in feet per second.

miles per hour

speed

kilometers per hour

feet per second

Try It

Things can be measured in many different ways.

Look at the list below. Put a check mark next to the main idea of the list. This is the one thing that all the other things have in common.

______ tons

______ pounds

______ weight

______ ounces

The main idea is *weight*. Look at the other words. They are all different kinds of weight. The weight of a truck is measured in *tons,* people's weight is measured in *pounds,* and the weight of a candy bar can be measured in *ounces*.

Put a check mark next to the main idea of this list.

______ time

______ seconds

______ hours

______ minutes

The main idea is *time. Seconds, years,* and *minutes* are all ways of measuring time.

Practice

Put a check mark next to the main idea in each list below.

1. ______ feet
   ______ yards
   ______ inches
   ______ height

2. ______ years
   ______ time
   ______ months
   ______ days

3. ______ liquid amounts
   ______ cups
   ______ quarts
   ______ teaspoons

4. ______ billions
   ______ thousands
   ______ numbers
   ______ millions

Check your answers on page 117.

Follow-Up

Most foods are sold by weight, for example, a pound of meat or 8 ounces of juice. Make a list of things you buy by the pound.

Lesson

16

Simple Machines

What You Know Have you ever noticed all the steps you follow when you brush your teeth? You (1) wet the toothbrush, (2) put toothpaste on the toothbrush, (3) brush your teeth, (4) rinse your mouth, (5) rinse the brush, and (6) put away the toothbrush and toothpaste. You probably have never thought about these steps or the order in which you do them. This is because the steps are easy and familiar to you.

When you do something you have never done before, you need to think about what you are doing. You also must think about what to do first, second, and so on.

How It Works

Some jobs around the house can't be done with only your hands or body. Sometimes you have to make a simple machine to help you. For example, you might have to move a very heavy TV set from a shelf on one side of the room to a shelf on the other side of the room.

Read the steps on the next page. They describe how to make a simple machine to help you do the job.

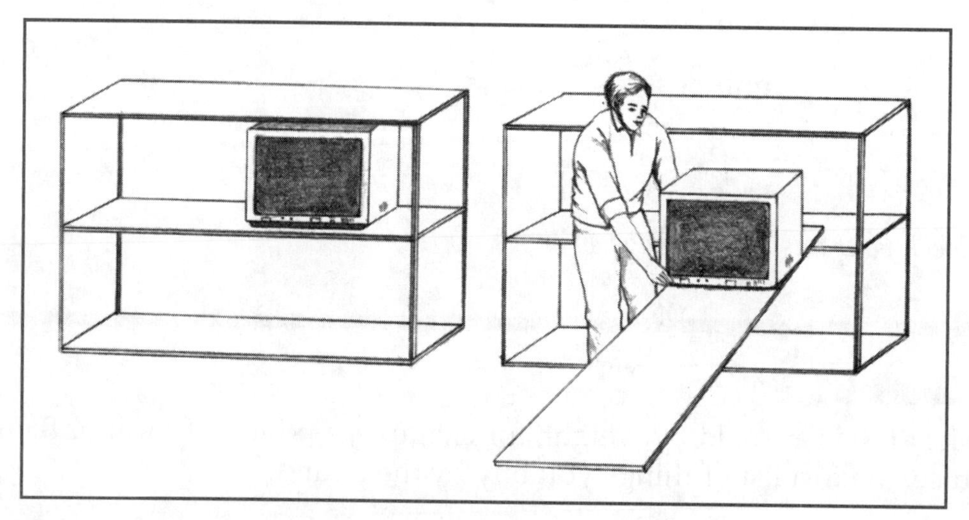

A board can become a simple machine.

1. Place a long, flat, strong board on the floor in front of the shelf where the TV is.

2. Raise one end of the board to the edge of the shelf the TV is on.

3. Slide the TV to the edge of the board.

4. Make sure that the board is resting solidly on the floor and the shelf.

5. Slowly slide the TV down the board to the floor.

6. Push the TV across the floor to the other side of the room.

7. Move the board in front of the other shelf and place one end on the edge of the shelf, with the other end on the floor, as you did before.

8. Turn the TV around so that it is facing outward.

9. Slide the TV up the board and onto the shelf.

Try It

Sometimes a drawer gets too full. When this happens, the drawer can get stuck, and you can't open it. You can make a simple machine to help you open the drawer.

Look at the list of steps on page 52. Decide what you would do first, second, third, and fourth by putting the numbers 1, 2, 3, or 4 in front of each step.

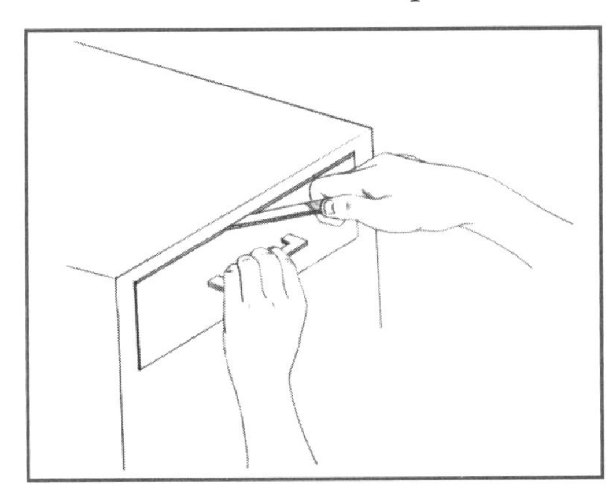

A letter opener can be used as a simple machine.

_____ Find what is blocking the drawer by moving the letter opener around.

_____ Slide the drawer open.

_____ Use the letter opener to push down whatever is blocking the drawer.

_____ Slip a letter opener into the top of the drawer.

Your list of numbers should read 2, 4, 3, and 1. The first step is to slip a letter opener into the top of the drawer. You need to use a letter opener because your fingers are too thick. The letter opener is a simple machine. It does something you can't do with your own hands.

Then you move the letter opener around until you find what is blocking the drawer. After that, you can use the letter opener to push down whatever is blocking the drawer. Finally, you can slide the drawer open.

Practice 1

Sometimes the pages of a book get stuck together in several places. You can use a simple machine to help you unstick the pages without tearing them.

Read the following steps. Show what you would do first, second, third, fourth, and fifth by writing numbers (1, 2, 3, and so on) in front of each step.

How to Unstick the Pages of a Book

_____ Repeat the process for other pairs of stuck pages.

_____ Carefully move the letter opener back and forth, hitting the stuck part gently each time.

_____ Find two pages that are stuck together.

_____ Insert a letter opener between the two stuck pages.

_____ When the pages become unstuck, open the pages out flat.

Practice 2

Some drinks taste better when they have crushed ice in them. You can make a simple machine to help you crush ice.

Read the following steps. Show what you would do first, second, third, and fourth by writing numbers (1, 2, 3, and 4) in front of each step.

How to Make Crushed Ice

______ Use a hammer or rolling pin to hit the plastic bag quickly and hard a number of times.

______ Put the pieces of crushed ice into a bowl.

______ Put the ice cubes into a plastic bag.

______ Remove ice cubes from the ice-cube tray.

______ Place the plastic bag on a flat, hard surface.

Check your answers on page 117.

Follow-Up

We make or use simple machines all the time. What kinds of simple machines have you used to help you do a job in the past week? Write four or five sentences explaining how you used one simple machine.

Lesson 17

The Laws of Motion

What You Know When a baby cries, you know that there is usually a reason. For example, the baby may be hungry, or he or she may have a wet diaper. By offering the baby some food and checking the diaper, you can usually figure out what caused the crying.

How It Works

One thing often leads to another. *Why* something happens is called the *cause*. *What* happens is called the *effect*. If you push very hard on one side of a refrigerator, it will begin to slide in the direction you are pushing. Your pushing is the *cause* (*why* it moves) and the refrigerator's movement is the *effect* (*what* happens).

Everything in the world is controlled by the laws of motion. One law of motion is called the law of **inertia** (ihn-ER-shuh). This law has two parts.

- One part of the law says that when something is standing still, it will stay that way unless something causes it to move.
- The other part of the law says that when something is moving, it will keep moving unless something stops it.

Have you ever tried to push a stalled car? At first, it's very hard to get it to move. Then, once it's rolling along, you have to step on the brake to stop it.

This example shows both parts of the law of inertia. It's hard to get the car to move at first because it's standing still. When the car is standing still, it will stay that way unless something causes it to move. This is the first part of the law of inertia.

When you finally get the car to move, it is hard to get it to stop. When the car is moving, it will continue moving unless something stops it. This is the second part of the law of inertia.

Try It

Put a check mark next to the *cause* that most likely leads to the *effect*.

How far can a dime roll?

Effect:
A dime rolls 50 feet down a sidewalk.

Cause:

______ The dime was lighter than most dimes.

______ Once the dime got moving, nothing stopped it.

______ It was a brand-new dime.

The dime kept rolling for such a long way because, once it got moving, nothing got in the way to stop it. This illustrates the second part of the law of inertia. Once something starts moving, it will keep on moving until something causes it to stop.

The first answer is wrong. All dimes are the same weight. There are no "light" dimes. The third answer is also wrong. How new a dime is has nothing to do with how far it will roll.

Practice

Read each sentence below. Then put a check mark next to the *cause* that leads to the *effect*.

1. **Effect:** A small stone rolls along a path in the woods.

   Cause: ______ A bear nearby makes a noise.

   ______ A bird sits on the stone.

   ______ As a deer walks by, it kicks the stone.

2. Effect: You step on the brake, but the car doesn't stop for several seconds.

Cause: ______ Inertia keeps the car moving.

______ Inertia helps stop the car.

______ You stepped on the brakes too hard.

3. Effect: You push on the end of a dresser, but it won't move at first.

Cause: ______ The dresser is made of wood.

______ Inertia keeps the dresser from moving.

______ The dresser is three inches from the wall.

Check your answers on page 117.

Follow-Up

According to the first part of the law of inertia, when something is standing still, it will stay that way until something moves it. Write down two things that have happened to you that are examples of this.

According to the second part of the law of inertia, when something is moving, it will keep moving until something stops it. Write down two examples of this that have happened to you.

How Heat Moves

What You Know Have you ever bought something and not been able to figure out how it works? If you have, then you know that written instructions are not always clear or helpful. It's a good thing that instructions often come with pictures. These pictures often have arrows pointing at things and words explaining what you are looking at.

How It Works

Pictures that explain how something works are called **diagrams** (DEYE-uh-grams). They are simple drawings that may not look exactly like the real thing. Diagrams often have words, arrows, and other information on them. Here are some steps to follow when you read diagrams, so that you will be able to make good use of them:

1. Look at the title of the diagram. It tells exactly what the diagram shows.

2. Before studying the diagram, read the written material that goes with it. See how much you can understand without looking at the diagram.

3. Look at the different parts of the diagram. Figure out what each part is and how all the parts fit together.

4. Look at the words (which are usually called labels), arrows, and other information on the diagram. What do they tell you?

5. Picture in your mind what is happening in the diagram. What happens first? What happens next?

6. Finally, go over the written material again. Compare what you are reading with what you see in the diagram.

Try It

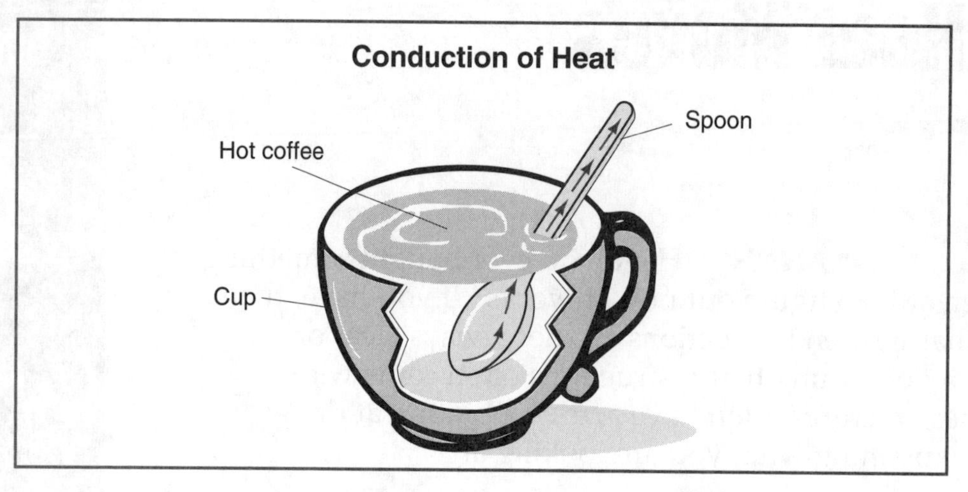

This is a diagram of the conduction of heat.

Look at the diagram above and read the explanation below. Then answer the questions.

When you put a spoon into hot coffee, the handle of the spoon doesn't touch the coffee at all but the handle still gets hot. The hot coffee heats the bottom of the spoon. Then the heat moves up the spoon to the handle. This movement of heat through a solid object is called **conduction** (kuhn-DUHK-shuhn).

1. What is the title of this diagram?

____

The title of the diagram is "Conduction of Heat."

2. What two solid things are shown in the diagram?

__________ __________

The diagram shows a spoon in a cup.

3. What labels are on the diagram?

__________ __________ __________

The words *cup, hot coffee,* and *spoon* are on the diagram.

4. What do the arrows on the spoon mean?

____

The arrows show that something moves up the spoon.

5. What happens first? What happens next?

____

____

First the coffee makes the bottom of the spoon hot. Then the heat travels up the spoon.

6. What else does the explanation tell you?

____

The explanation says that the movement of heat through a solid object is called *conduction.*

Practice

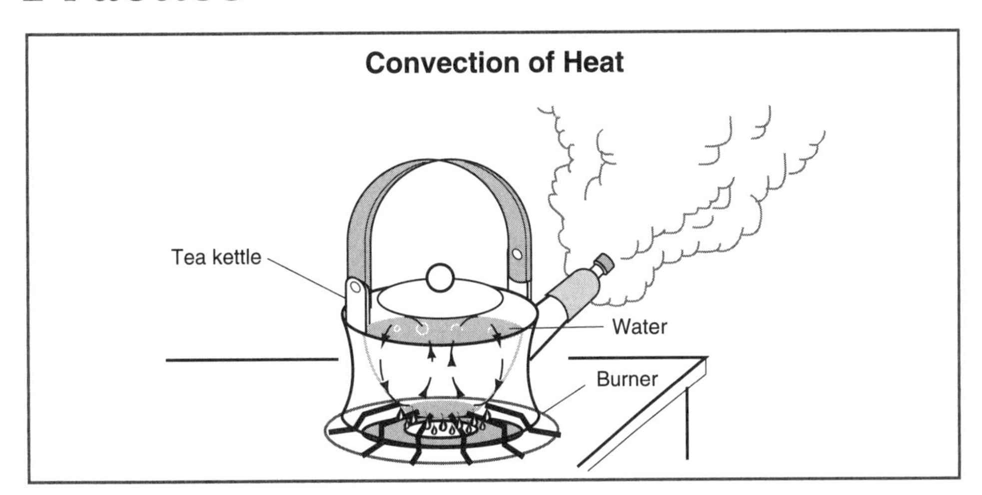

This diagram shows the convection of heat.

Look at the diagram above and read the explanation below. Then answer the questions.

When you heat water on the stove, the hottest part of the water is at the bottom of the kettle. The water at the top is a little cooler. The hot water rises. This pushes the cooler water down to the bottom, where it gets warmer. After a while, all the water in the kettle is hot. This kind of movement of heat through a liquid is called **convection** (kuhn-VEK-shuhn).

1. What is the title of this diagram?

____

2. What two solid things are shown in the diagram?

______ ______

3. What labels are on the diagram?

______ ______ ______

4. What do the arrows on the water show?

______

______

5. What happens first? What happens next?

______

______

6. What else does the explanation tell you?

______

______

Check your answers on page 117.

Follow-Up

On a very hot, sunny day, a car hood is almost too hot to touch. However, the wood on the tree next to the car is much cooler. Can you guess why this is? Do you think it has something to do with conduction or convection of heat?

Unit 4

Energy

Lesson 19

Atomic Energy

What You Know When people get excited, they sometimes mistake feelings for facts. A child who says "You have to buy me a bicycle. I need it!" may think that this statement is a fact. However, you know the facts. The bicycle costs $200. The store is closed. "I need this bicycle" is a statement of feeling, not fact.

How It Works

Atomic (uh-TAHM-ihk) **energy** is power that comes from very small things called **atoms** (AT-uhmz). Sometimes atomic energy is called **nuclear** (NOO-klee-uhr) **energy**. The energy can be used to make electricity.

Some people think that atomic energy is good because it makes electricity without putting poisons in the air the way other kinds of energy do.

Other people think it is too dangerous. If there is an accident at a nuclear power plant, the materials that make the energy may get into the air. These materials can make people very sick or even kill them.

The Palo Verde Nuclear Power Plant is shown in this photo.

In Lesson 14 you read about the difference between facts and opinions. There are a few ways for you to tell the difference.

Facts

You can count, measure, or see facts.

- There are two nuclear power plants located in the state of Ohio.
- The Fermi 2 Power Plant in Ohio started making electricity in 1988.

Facts can be proven to be real or true. They do not change. The *1995 Information Please Almanac* says that the South Texas 1 Nuclear Power Plant is the largest nuclear power plant in the United States.

Opinions

Opinions cannot be proven to be true. They are based on personal values or points of views. One way you can tell that something is an opinion is to look for key words such as *good, better, best, safe, should,* and *dangerous.*

- Nuclear power plants are really dangerous.

Opinions may also be based on feelings or beliefs. These opinions often start with *I think* or *don't think, I feel, I believe,* or _______ *should.*

- I think that all nuclear power plants should be shut down right away.

It often happens that there are many different opinions about the same thing.

- The government should not allow more nuclear power plants to be built.
- The government should help states build more nuclear power plants.

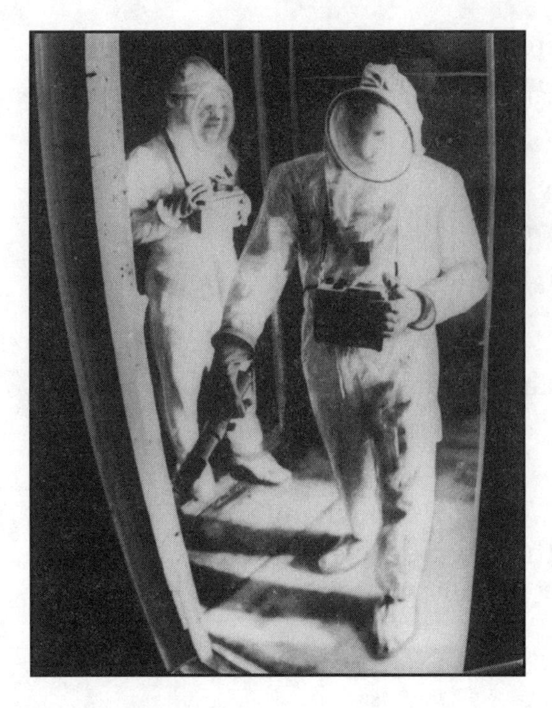

After there is an accident at a nuclear power plant, people must wear special suits to protect them until they are sure the air is safe.

Try It

Write F in the answer blank if the sentence is a fact. Write O if the sentence is an opinion.

___ **1.** There was an accident in a nuclear power plant in Chernobyl, (chuhr-NOH-buhl), in the former Soviet Union, in 1986.

Fact. Scientists visited the plant and saw what happened. Newspaper reporters visited hospitals and saw sick people.

___ **2.** I think that an accident like the one at Chernobyl can happen again.

Opinion. This statement is based on a belief. Notice the words *I think*.

___ **3.** In the history of nuclear energy, there have been six accidents at nuclear power plants around the world.

Fact. Facts can be proven. They do not change. Written records show the number of accidents.

Practice

Read the following sentences. Write F in the answer blank if the sentence is a fact. Write O if the sentence is an opinion.

___ **1.** Nuclear power plants cost more to build than regular power plants.

___ **2.** Nuclear power plants cost less to operate than regular power plants.

___ **3.** I don't think that we should spend tax dollars on nuclear power plants.

___ **4.** Europe gets 50 percent of its electrical power from nuclear power plants.

___ **5.** Europe has the best electrical power system in the world.

___ **6.** The Nuclear Energy Commission (a group formed by the federal government) controls all nuclear power plants in the United States.

___ **7.** Nuclear energy may be clean, but I don't want a power plant near where I live.

Check your answers on page 117.

Follow-Up

List other kinds of energy you know about. (Hints: One is up in the sky. Another is something you drink.)

Saving Energy at Home

What You Know When you first receive a letter from a friend, you may not have time to read the whole letter because you may be on your way to work. You may look at it quickly and then put it away to read later. The first time through, you probably just look for the important points or special things that are of interest to you.

How It Works

In Lesson 15 you learned about finding the main idea in a list. In this lesson, you will learn how to find the main idea in a group of sentences.

To find the main idea, follow these steps:

1. First decide in a general way *what* or *who* all the sentences are about. Read these sentences.

   Turn off the lights when you leave a room.

   Don't keep the refrigerator door open too long.

   You can save electricity in several ways.

   Use 60-watt bulbs instead of 100-watt bulbs.

   What are these sentences about? They are about saving electricity.

2. Now find the most important and the most general point that is made about saving electricity. It is, *You can save electricity in several ways.*

3. Look at the other three sentences. Each one is a specific way of saving electricity. These are the *details* that support, or help explain, the main idea.

Try It

1. Read the sentences below. Put a check mark next to the one that contains the main idea.

   ______ Thick curtains help block the sun's heat.

   ______ Use an electric fan to move the air.

   ______ You can stay cool in the summer without using an air conditioner.

   ______ Eat foods that don't have to be cooked.

   All the sentences are about staying cool in the summer. The most important point is in the third sentence: *You can stay cool in summer without using an air conditioner.* The other three sentences are details that explain how to do this. They are details that support or help explain, the main idea.

2. Now read these sentences. Put a check mark next to the one that contains the main idea.

   ______ Turn the temperature down to 68 degrees.

   ______ Small changes can help you stay warm and save energy in the winter.

   ______ Wear a sweater when it gets cold.

   ______ Open the curtains on windows that receive direct sunlight.

   All the sentences are about staying warm without wasting energy. The main idea is contained in the second sentence: *Small changes can help you stay warm and save energy in the winter.* The other three sentences give details about how to do this.

Practice

Adobe houses have thick walls.

Put a check mark next to the sentence that contains the main idea.

1. ______ **Adobe** (uh-DOH-bee) houses never get too hot or too cold inside.

   ______ The walls of adobe houses soak up heat during the day.

   ______ The walls of adobe houses give off heat when it's cool at night.

   ______ Adobe houses stay cool inside during the day.

2. ______ Adobe bricks are made of soil, straw, and water.

   ______ Adobe bricks bake in the sun for 7 to 14 days.

   ______ Adobe bricks are easy to shape when they're wet.

   ______ Adobe bricks are cheap and easy to make.

Check your answers on page 118.

Follow-Up

Write three supporting sentences for the following sentence, which is the main idea: I do several things to save energy at home.

Solar Energy

What You Know When you open the refrigerator door and the light doesn't go on, you know there must be a reason. The light bulb may have burned out, or the refrigerator may be unplugged. By checking the plug and putting in a new bulb, you can usually figure out what caused the problem.

How It Works

In Lesson 17 you learned about *cause* and *effect*. An event that makes something else happen is a *cause*. This is *why* something happened. The result of a cause is the *effect*. This is *what* happened.

You can use cause and effect to understand more about **solar** (SOH-luhr) **energy**, which is energy from the sun. People are beginning to use solar energy to make hot water at home, to help heat their homes, and even to charge batteries.

Put a check mark next to the effect that resulted from the cause.

Cause: The sun shines on a car whose windows are closed.

Effect: ______ The air in the car gets cooler.

______ The air in the car gets warmer.

______ The air in the car stays the same temperature.

The air in the car gets warmer. The sunlight shines through the windows into the car. The air in the car is heated from the sunlight. There is no way for the hot air to escape and there is no way for cooler air to get into the car. So *what* happens? The air in the car gets warmer.

Try It

Some homes already have solar collectors. These collect the heat of the sun and turn it into energy.

Put a check mark on the lines below next to the effect that resulted from the cause.

Cause: Supplies of fuels like gas and oil are growing smaller.

Effect: ______ Everyone stops using gas and oil to heat their homes.

______ Some people are using solar energy to heat their homes.

______ People stop heating their homes.

The correct answer is the second one: *Some people are using solar energy to heat their homes*. This is *what* happens as a result of smaller supplies of other fuels. People do not stop using gas and oil to heat their homes. There is still gas and oil left. It takes time to find new sources of energy. People cannot stop heating their homes entirely. In cold climates they would be very uncomfortable and probably would get sick if they didn't heat their houses.

Practice

Put a check mark next to the *effect* that resulted from the *cause*.

1. Cause: Dark glass takes in more heat than clear glass.

Effect: ______ Solar collectors use clear glass.

______ Solar collectors use black glass.

______ Solar collectors use yellow glass.

2. Cause: The hottest sunlight comes from the south.

Effect: ______ Solar collectors face south.

______ Solar collectors face north.

______ Solar collectors are square.

3. Cause: Solar collectors work only on sunny days.

Effect: ______ They work well in rainy climates.

______ They work well at night.

______ They don't give a steady supply of energy.

Check your answers on page 118.

Follow-Up

What do you think are the advantages of solar energy? List two. What are the disadvantages? List two of these as well.

Two Kinds of Energy

What You Know When deciding what to do on a Saturday night, you might think of several different possibilities. You could go dancing, see a movie, visit a friend, watch TV, or go food shopping. Some of these activities are fun, and some aren't. Some cost money, and some don't. You compare the possibilities before you decide what to do.

How It Works

Scientists say that energy is the ability to "do work." If something contains energy, it can move something or heat something up. For example, people have energy. The gas in your stove has energy.

There are two types of energy, **potential** (poh-TEHN-shuhl) **energy** and **kinetic** (kih-NEHT-ihk) **energy**. In some ways they are alike, and in some ways they are different.

When you **compare** (kuhm-PAIR) two things, you tell how they are the same. When you **contrast** (kuhn-TRAST) two things, you tell how they are different.

First, let's compare the two types of energy.

Things that contain kinetic energy and potential energy both have the ability to do work. This is how they are the same.

Now let's contrast the two types of energy.

- Potential energy is stored energy. If an object contains potential energy, it can do work in the future.

For example, a pile of wood has potential energy. It can make heat in the future, but it isn't making heat right now. A sleeping athlete also has potential energy. He or she can move but is not moving right now.

- Kinetic energy is active energy. A pile of wood that is burning has kinetic energy. The burning wood is creating heat. The potential energy stored in the wood has turned into kinetic energy. The athlete who is running a race also has kinetic energy.

Try It

Look at the pictures and place a check mark next to the correct comparison or contrast below each one.

What kind of energy does each of these have?

1. ______ The battery has kinetic energy, but the candle doesn't.

   ______ The candle has kinetic energy, but the battery doesn't.

   ______ Both the battery and the candle have kinetic energy.

The second answer is correct: *The candle has kinetic energy, but the battery doesn't.* The candle is "doing work" now. It is burning and giving off heat. The first and third answers are wrong because the battery doesn't have kinetic energy. It isn't "doing work" right now.

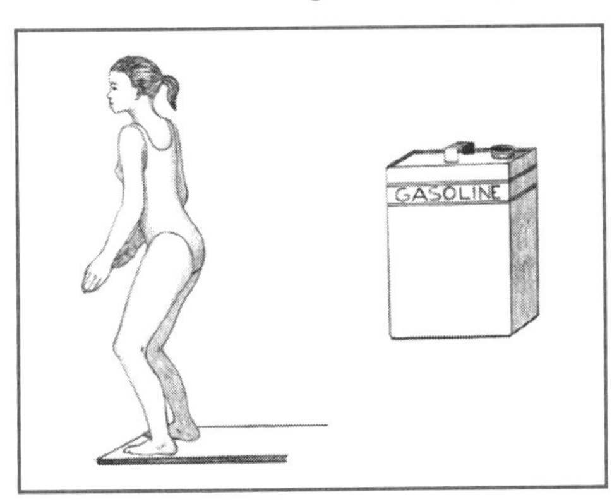

What kind of energy does each of these have?

2. ______ Both the diver and gasoline have potential energy.

   ______ Both the diver and gasoline have kinetic energy.

   ______ The diver has kinetic energy but the gasoline doesn't.

The first answer is correct: *Both the diver and gasoline have potential energy.* They both contain stored energy, but neither is "doing work" now.

Practice

Look at the pictures and place a check mark next to the correct comparison or contrast.

What kind of energy does each of these have?

1. ______ The firecracker has kinetic energy, but the football player doesn't.

   ______ The football player has kinetic energy, but the firecracker doesn't.

   ______ Both the firecracker and the football player have kinetic energy.

2. ______ The football player has potential energy, but the firecracker doesn't.

   ______ The firecracker has kinetic energy, but the football player doesn't.

   ______ The football player and the bird both have kinetic energy.

Check your answers on page 118.

Follow-Up

Make a list of things in your home that have either potential energy or kinetic energy.

Different Energy Sources

Lesson 23

What You Know

Do you ever watch the people coming out of a movie? They may be smiling and laughing, or they may be looking sad. Their faces "say something" about the movie they just saw. You can guess what they think without their telling you in words.

How It Works

When you read, you will find that certain facts, information, ideas, and feelings are not written in the words. When this happens, you must **infer** (ihn-FER) the information that is not written in the words.

Making an **inference** (IHN-fuhr-ihns) means that you figure out what the unwritten information is. You do this by using the information that is written plus what you know from your own life. You put these two things together to figure out what is not written.

For example, if you see someone wearing tennis shoes and carrying a tennis racket, you can *infer* that the person will play, or has just played, tennis. If you see someone leading 15 children down the street, you can infer that the person is a teacher or child care worker.

Read the following sentences. What can you infer?

> We are running out of supplies of coal, oil, and natural gas. It will take millions of years to make a new supply. We will have to change the way we heat our homes.

Can you infer that in the future we will not heat our homes with coal, oil, or natural gas? (Yes. There won't be any coal, oil, or natural gas left.)

Can you infer that we will heat our homes with wood? (No. There is no information about what we will use to heat our homes in the future.)

Try It

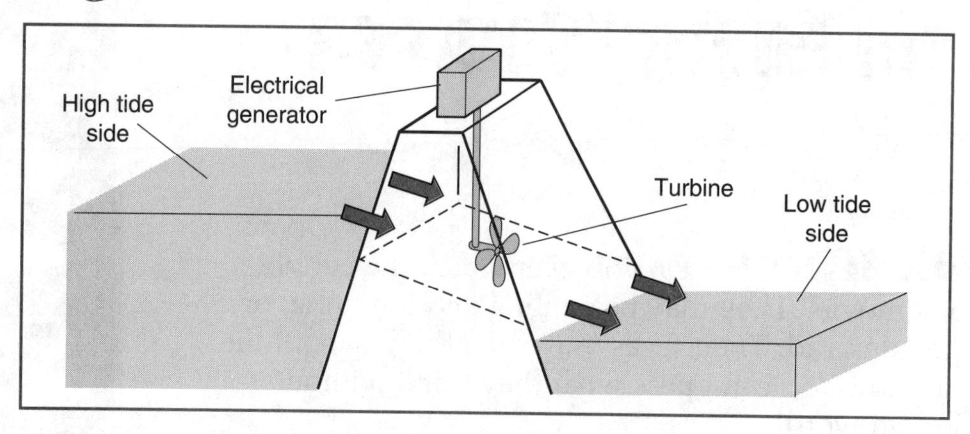

This drawing shows a tidal energy station.

Read the following sentences. What can you infer from them?

> **Tidal** (TEYED-uhl) **energy stations** can be used to make electrical power. As the ocean tide comes in, the water flows around **turbines** (TER-bihnz) and makes electricity. (A turbine is a kind of engine.) This is a cheap, clean way to make electricity. However, many plants and animals living near these kinds of power stations are hurt by them.

Read the following sentences. Write Yes on the answer blank if you have enough information to make the inference. Write No if you do not.

___ **1.** There will never be a tidal energy station in Kansas.

Yes. Look at the map of the United States in Lesson 12 to see where Kansas is. The paragraph above says that tidal energy stations use the ocean's tides to make electricity. Kansas is not near the ocean, so you can infer that there will never be a tidal energy station in Kansas.

___ **2.** The number of fish near a tidal energy station decreases each year.

Yes. The story says that tidal energy stations hurt nearby plants and animals. Fish are animals that live in the ocean. You can infer that fewer fish would survive near a tidal energy station.

___ **3.** There are more tidal energy stations than nuclear power plants in the United States.

No. The story doesn't say anything about how many tidal energy stations there are in the United States.

Practice

Read the following sentences.

Iceland uses mainly water power. Almost all of its electrical power is produced by **hydroelectric** (HEYE-droh-ee-LEHK-trihk) **stations**. These plants use the power of waterfalls to make electricity. Also, water from hot springs is used to heat homes, offices, and some industries. **Hot springs** are waters that are warmed underground by nature.

Some waterfalls on Iceland's rivers are used for power to make electricity.

Read the following sentences. Write Yes on the answer blank if you have enough information to make the inference. Write No if you do not.

___ **1.** Iceland won't build nuclear power plants in the future.

___ **2.** Iceland is very cold.

___ **3.** People have a lot of electrical appliances in Iceland.

___ **4.** Iceland has many rivers.

___ **5.** Some of Iceland's electrical power comes from wind power.

Check your answers on page 118.

Follow-Up

In Lesson 12, you learned that some kinds of animals are dying out because people have taken over the land where they lived. In this lesson, you have learned that we are running out of coal, oil, and natural gas. Write two or three sentences telling what you can infer from these two things.

Lesson 24

What Fuels Do We Use?

What We Know You probably look at a clock many times every day. The numbers 1 through 12 divide the face of a clock into 12 equal sections, like pieces of a pie. When you think of a clock, it's easy to see that 3 hours is one quarter of 12 hours.

How It Works

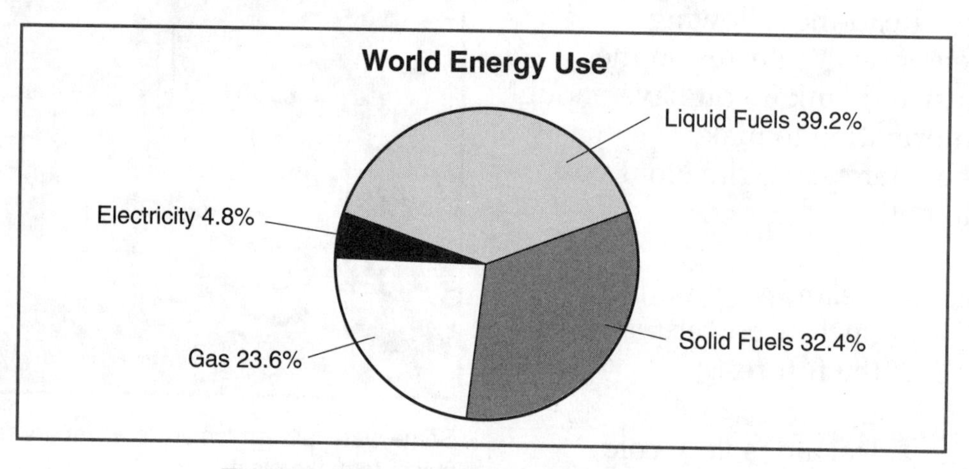

This is a pie chart showing world energy use.

The circle above is called a **pie chart**. It is a chart in circle form. It is divided into several pieces, just like a pie. However, the pieces of this chart are not all the same size. A pie chart gives you a clear picture of the parts that make up a whole thing. It also shows how big each part is when compared with the others.

Look at the title of this chart. It tells you what the chart shows. This pie chart title is "World Energy Use." The chart shows how much of each type of fuel we use. *Solid fuels* are things like coal and wood. On this chart, *gas* doesn't mean "gasoline." It means the kind of gas that comes out of a gas stove when you turn it on. *Liquid fuels* include the kind of gas you use in your car along with heating oil.

The chart also gives percentages for the amount of each type of fuel used. For example, the chart shows that only 4.8 percent of the fuel we use is in the form of electricity. Most of our energy comes from solid, liquid, and gas fuels.

Try It

Look at the pie chart on page 78. Put a check mark next to the correct answer.

1. Which type of fuel do we use the most?

______ A. Solid fuels

______ B. Liquid fuels

______ C. Gas

The answer is B. We use liquid fuels the most.

2. Which type of fuel do we use the least?

______ A. Liquid fuels

______ B. Electricity

______ C. Gas

The answer is B. Electricity is used the least.

3. What percent of the total amount are solid fuels?

______ A. 39.2 percent

______ B. 23.6 percent

______ C. 32.4 percent

The answer is C. Solid fuels are 32.4 percent of the total.

4. Which type of fuel accounts for 39.2 percent of the total fuel used?

______ A. Liquid fuels

______ B. Electricity

______ C. Gas

The answer is A. Liquid fuels are 39.2 percent of the total.

Practice

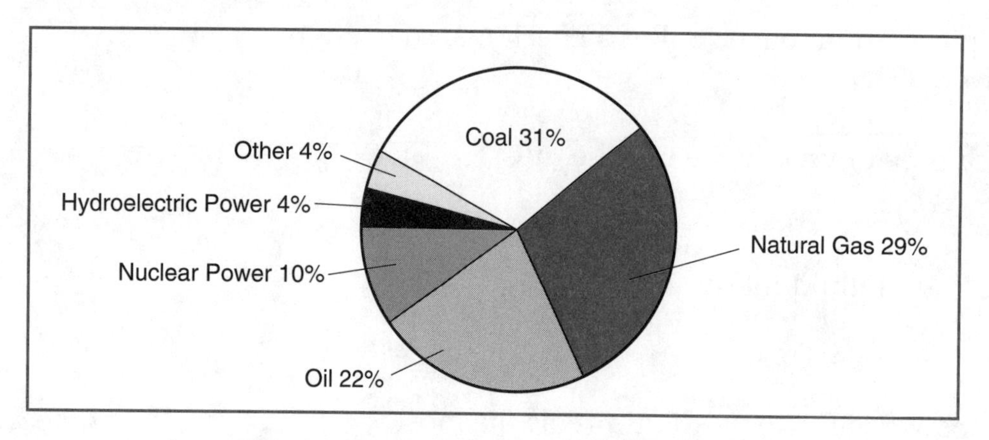

This pie chart shows U.S. energy production.

Look at the chart above. Put a check mark next to the correct answer for each question.

1. What type of fuel does the United States produce the most?

______ A. Coal

______ B. Oil

______ C. Natural gas

2. What percent of the total amount is oil?

______ A. 4 percent

______ B. 10 percent

______ C. 22 percent

3. What fuel accounts for 31 percent of the total?

______ A. Hydroelectric power

______ B. Nuclear power

______ C. Coal

Check your answers on page 118.

Follow-Up

What kind of fuel is used to heat your home? Run the stove? Heat the water? Ask two of your friends these questions and see which kind of fuel is used for each different purpose.

Unit 5

The Solar System

How the Earth Moves

What You Know Suppose you are watching a weather report on TV and the telephone rings. While you are talking on the phone, you miss part of the report. All you hear is, "It will be . . . and warmer tomorrow with . . . in the 70s. No rain in sight." By thinking about the words you did hear, you can guess that the words you missed were *sunny* and *temperatures*.

How It Works

You can't always understand every word or sentence you read. However, you can learn to make good guesses. In Lesson 9 you practiced guessing the meaning of new words by looking at nearby words and sentences. Now read the following paragraph.

> People have been studying the skies since before they invented ways to read or write. In the Middle East, early astronomers (uh-STRAHN-uh-muhrz) looked at the movements of the moon and of the planets, including Earth, as they tried to make calendars. These early scientists came up with several different explanations for the changes they saw in the night sky. Although some of their ideas were wrong, these astronomers had ideas that led to modern ideas about how the Earth and the moon move.

What does the word *astronomer* mean? Look at each sentence in the paragraph. The first sentence mentions people who study the sky. The second sentence says that the early astronomers *looked at the movements of the moon and of the planets, including Earth*. The third sentence talks about *early scientists*. The fourth sentence says astronomers had ideas about *how the Earth and the moon move*.

By putting together the information in these sentences, you can figure out that an *astronomer* is a scientist who studies the movements of the moon and of planets, including Earth.

Try It

This is a drawing of what an early Chinese observatory might have looked like.

About 1300 BC (before Christ) the Chinese started building observatories (uhb-ZERV-uh-toh-reez). These places were little more than high platforms from which they could get a clear view of the sky. Early scientists studied many celestial (suh-LEHS-chuhl) happenings. These included the rising and setting of the sun and the changes in the shape of the moon.

1. What were the first *observatories*?

______ A. scientists who first studied stars

______ B. the first places from which people could look at the sky

______ C. the first changes of the moon that people studied

The correct answer is B. The second sentence says that observatories were built as places from which people could get a clear view of the sky.

2. What does *celestial* mean?

______ A. relating to the night

______ B. relating to the sky

______ C. relating to events

The correct answer is B. The last two sentences say that the scientists studied the sun (not just the night sky), and not all events, just those in the sky.

Practice

Look at the diagram below. Then read the paragraph. Put a check mark next to the correct answer to each question.

The earth revolves around the sun once every 365¼ days.

The work of early astronomers contained some **misconceptions** (mihs-kuhn-SEHP-shuhnz). For example, some early astronomers thought that the earth was standing still in space. As time passed, we learned the truth—that the earth moves around the sun in a curved path once every 365¼ days. We now also know that the moon **revolves** (rih-VAHLVZ) around the earth, just the way the earth revolves around the sun.

1. What are *misconceptions*?

___ A. good pictures ___ B. large planets ___ C. wrong ideas

2. What does *revolve* mean?

___ A. move in a curved path ___ B. become larger ___ C. go away

Check your answers on page 118.

Follow-Up

Look up at the night sky every night for about a week. Write the date and make a drawing each night you see the moon. What changes do you see in the shape of the moon?

Reasons for Seasons

What You Know If you are late to work one day, your boss may want to know why. He or she is probably not interested in all the details. (There was no electricity in your neighborhood, so the traffic lights weren't working. Because of this, the bus you were on had an accident. It took a long time for another bus to come.) Your boss just wants a quick explanation. (The bus I was on had an accident.)

How It Works

In Lesson 15 you learned how to find the main idea of a list of words. In Lesson 20 you learned how to find the main idea of a group of sentences. In this lesson, you will learn how to find the main idea of a paragraph.

You should treat a paragraph just like a group of sentences. First decide what all the sentences are about. What the sentences are about is called the **topic**. Then find the most important or the most general point about the topic. This is called the **main idea**. All the other sentences should contain details that support the main idea.

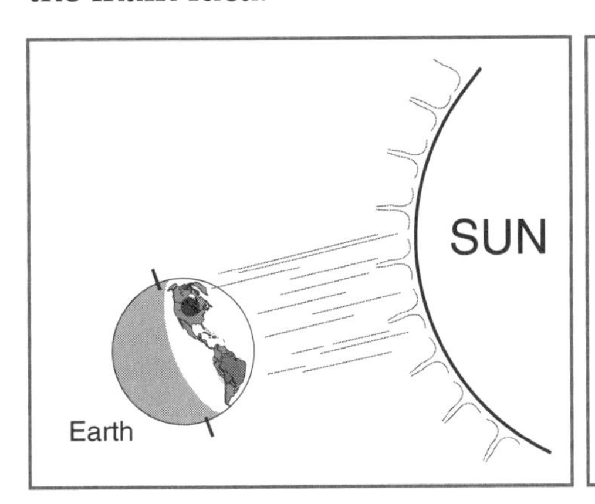

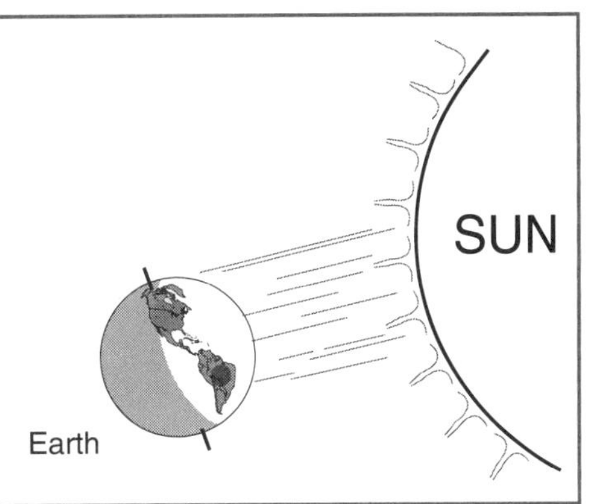

The drawing at the left shows where the sun is in the sky when it's summer in North America. The drawing at the right shows where the sun is when it's summer in South America.

Read the following paragraph.

> It is summer in the United States in June, July, August, and September. In the summer, the northern half of the earth is tipped toward the sun. This is when the sun's rays hit the United States more directly and cover a smaller area than at other times of the year. Because of this, more heat reaches the United States in the summer than in the winter. There are also more hours of sunlight in the summer.

What is the topic? It's *summer in the United States.*

Which sentence states the main idea of the paragraph? *It is summer in the United States in June, July, August, and September.* This sentence is the most important and the most general in the paragraph. All the other sentences contain details that support this main idea.

Try It

Underline the sentence that contains the main idea in the paragraph below.

> In early spring in North America, the earth is beginning to tilt toward the sun. The sun's rays are only beginning to strike North America more directly. In late fall, North America has already tilted quite far away from the sun. The sun's rays are now shining much less directly there. The weather in North America in early spring and late fall is quite similar.

What is the topic? *Spring and fall in North America.* What is the most important or most general sentence about spring and fall? *The weather in North America in early spring and late fall is quite similar.* This sentence contains the main idea. The other sentences have details that support the main idea.

Practice

Underline the sentence that contains the main idea in each of the paragraphs below.

1. The earth as a whole gets the same amount of energy from the sun each year. This energy comes in the form of heat and light. Seasons change because each part of the earth receives different amounts of the sun's energy at different times of the year. When the sun shines most directly on an area, this area has summer. When the sun shines least directly on an area, this area has winter.

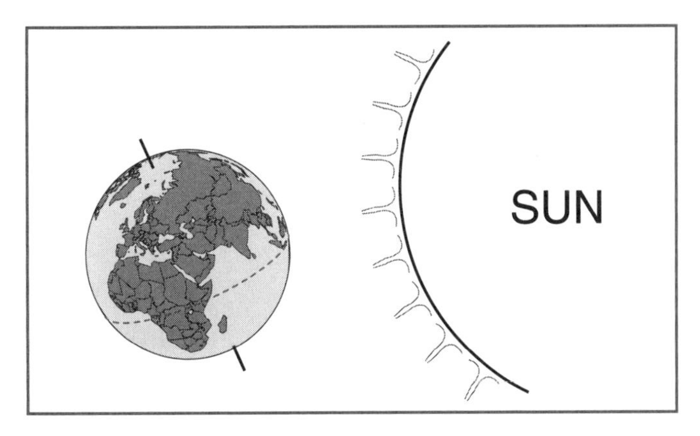

Countries on or near the equator get the direct rays of the sun all year round.

2. Some parts of the earth don't have seasons. In countries on or close to the equator, the fact that the earth tilts has little effect on the temperature. These countries have warm weather all year round. For example, during summer in South America, Ecuador receives the direct rays of the sun. When winter comes to South America, the sun's rays are still very direct on Ecuador, and the weather stays almost as warm as it is in summer.

Check your answers on page 118.

Follow-Up

Write a paragraph containing five sentences. The first sentence should start, "My favorite season is ____________."

Lesson

27

The Sun, Planets, and Moon

What You Know When you are shopping for clothes, it is often hard to decide what to buy. You may find more things you like than you can afford. For example, you may find two jackets that you like, but you can afford to buy only one. What do you do?

First you compare them to see what is similar about them. They both fit you, and they are both about the same price. Then you contrast the two jackets. What is different about them? One is tan, which is a color you can wear with a lot of your clothes. The other is plaid, which will make it hard to find a lot of things to wear with it, but you like this jacket a little more than the tan one.

Which one do you decide to buy? Of course, your answer will depend on what is more important to you—liking one jacket more or being able to wear the other jacket more. Comparing and contrasting can help you make up your mind.

How It Works

The *sun* is a **star**, just like all the other stars in the sky. It is made up entirely of gases. It is not solid anywhere. The sun makes heat and light. The temperature on the sun is about 11,000 degrees.

The earth is one of the **planets** (PLAN-ihtz) that revolve around the sun. Like all the other planets, the earth is round in shape. The earth is covered mostly with water. However, the top layer of the earth is solid—including the parts that are land and the parts that are under the water. The earth has air, which makes it possible for animals and plants to live.

The moon is a **satellite** (SAT-uh-leyet) of the earth. This means that it revolves around the earth. The moon is also round in shape. It is completely solid and has no air.

First **compare** the sun, earth, and moon. This means that you will look for ways in which they are alike. For example:

The sun, earth, and moon are round.

Then you **contrast** the sun, earth, and moon.

The sun is made up of gases. The earth and moon are solid.

Try It

Read the sentences. Put a check next to the correct answer.

1. How are the moon and sun similar?

______ A. They are both made up of gases.

______ B. Nothing could live on the sun or the moon.

______ C. They are both satellites of the earth.

The correct answer is B. Nothing could live on the sun because it is so hot and is made up of gases. Nothing could live on the moon because it has no air. Answer A is wrong because the moon is solid. Answer C is wrong because the sun is not a satellite of the earth.

2. How are the earth and moon different from the sun?

______ A. The earth and moon revolve around another heavenly body, while the sun does not.

______ B. The earth and moon make their own heat and light. The sun gets heat and light from the Earth.

______ C. The earth and sun are round, while the moon is not.

The correct answer is A. The earth revolves around the sun, and the moon revolves around the earth. The sun does not revolve around anything. Answer B is wrong because it is the sun that makes heat and light, not the earth and moon. Answer C is wrong because the sun is not round.

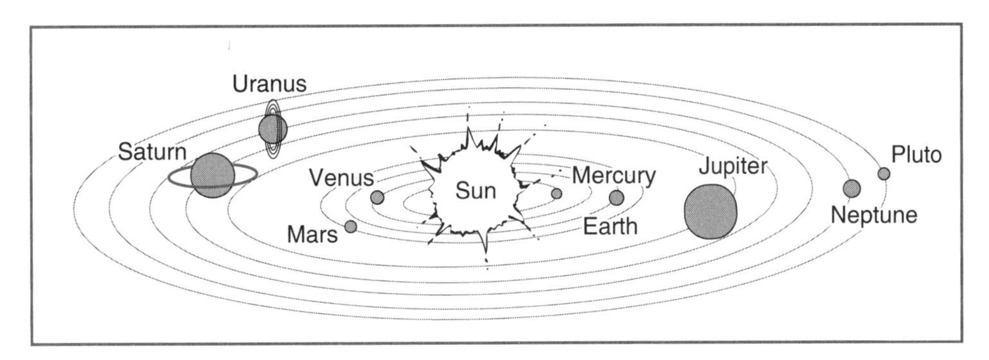

This shows the **solar** (SOH-luhr) **system**. *Solar* means relating to the sun. The solar system is made up of everything that is linked to the sun in some way. (This picture does not include any moons, which are also part of the solar system.)

Practice

Read the following paragraph. Then answer the questions by putting a check mark next to the correct answer.

> The nine planets in the solar system revolve around the sun. Each planet is round. Some planets have one or more moons; others have none. Some have rings around them. These rings are made up of small bits of ice that spin around the outside of the planet. Here is some information about two of the nine planets—Mars and Jupiter. Mars is the planet between Earth and Jupiter. Mars has two small moons and does not have any rings around it. Jupiter is the largest planet. It is almost three and a half times bigger than Mars and is more than five times the size of Earth. Jupiter has 16 moons and one ring around it.

1. How are Jupiter and Mars similar?

_____ A. Mars and Jupiter have a ring around them.

_____ B. Mars and Jupiter have more than one moon.

_____ C. Mars and Jupiter are smaller than the Earth.

2. How are Jupiter and Mars different?

_____ A. Mars is round; Jupiter is not.

_____ B. Mars has moons; Jupiter has none.

_____ C. Jupiter has a ring around it; Mars does not have a ring.

Check your answers on page 119.

Follow-Up

Do you think that there is life on other planets? Write a paragraph explaining why or why not.

The Constellations

What You Know When you shop for a new CD player, you may check the newspaper advertisements and visit different stores to see which store has the lowest prices. Then you buy your CD player at the store with the best price. Suppose a friend asks you where he or she should buy a TV. You would probably suggest the store where you bought your CD player. You can infer that the same store that has low CD player prices will have low TV prices, too.

How It Works

In Lesson 23 you learned to make inferences. You were able to figure out information that was not written. You did this by putting together what was written with what you know from your own life.

For example, you might see a sign that says, "Special Sale! This Week Only! All Bayway Autofocus Cameras $77.95." You can infer that the Bayway autofocus cameras usually cost more than $77.95.

Remember not to infer more than you should. When you see the advertisement, you might think that the cameras usually cost $89.95 or $99.95. But this may not be true. The usual price might be something like $79.95.

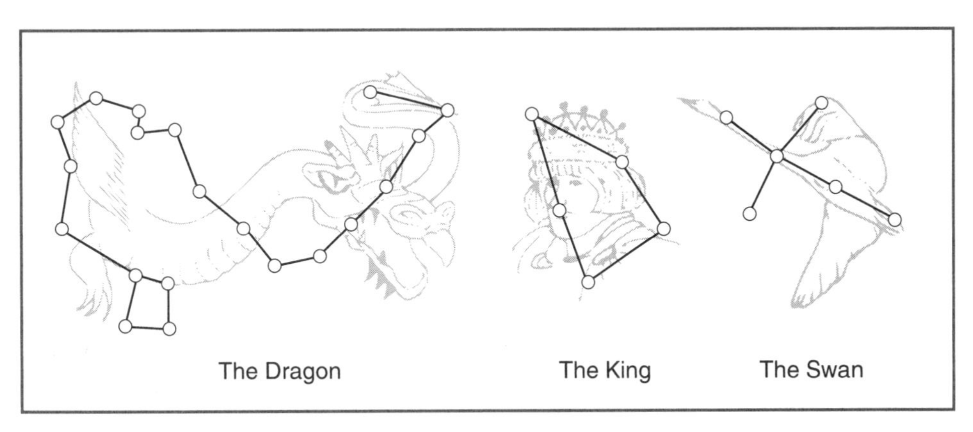

Three constellations are shown here. You will learn about constellations in this lesson.

Try It

Read the following paragraph. Then read the inferences that follow. Write Yes on the answer blank if you have enough information to make the inference. Write No if you do not.

> **Constellations** (kahn-stuh-LAY-shunz) are patterns of stars in the sky. If you connect the stars with lines as you do in a connect-the-dots game, some constellations form the shapes of animals and other things. At night, in the countryside, you can see up to 3,000 stars without even using a telescope. The bright lights of a city may make it impossible to see all but a few of the brightest stars.

___ **1.** Astronomers have named about 25 constellations so far.

No. The paragraph doesn't say anything about how many constellations there are or how many have been named.

___ **2.** You could not see 3,000 stars in the sky over New York City.

Yes. The paragraph says that in big cities you can see only a few stars.

___ **3.** You can see more than 3,000 stars at night in the countryside if you use a telescope.

Yes. The paragraph says that you can see up to 3,000 *without even using a telescope*. So you can infer that you could see more stars with a telescope.

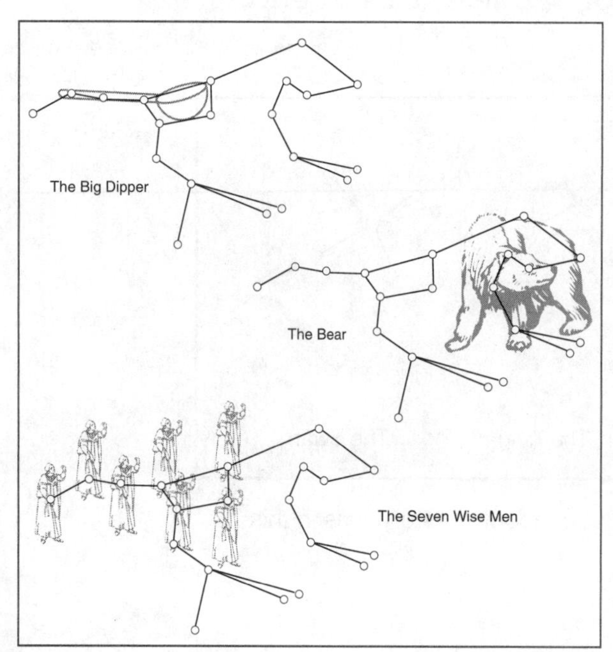

These three constellations are all made up of the same stars.

Practice

Read the following paragraph. Then read the inferences that follow. Write Yes on the answer blank if you have enough information to make the inference. Write No if you do not.

> One of the best-known constellations is called the *Big Dipper*. It is clearly visible in the night sky. It has been thought about and talked about by many different people since the beginning of time. This constellation is made up of seven bright stars. North Americans think that the stars form a shape like a water dipper (a ladle). In Greece, people see a bear shape. In India, people see seven wise men.

___ **1.** The stars look different in different countries.

___ **2.** Different people may see different things formed by the same stars.

___ **3.** People in Spain see the bear shape in these seven stars.

___ **4.** People in Greece and India don't see a water dipper because they don't have water dippers in those countries.

Check your answers on page 119.

Follow-Up

Look at the drawing on page 92. What other shapes can you find in this pattern of stars?

How Telescopes Work

What You Know The stories on soap operas continue from week to week and even from year to year. When people have a favorite soap opera, they want to keep up with the story. When they miss a day, they may ask a friend to tell them what happened. A show might take an hour to watch, but the friend can give the main idea in a minute: "Fran had her baby, Mario went back to Italy, and Lisa went out with Fred."

How It Works

In Lesson 26 you learned how to find the main idea in a paragraph. In this lesson, the main idea is not given in the paragraph. You have to decide what the main idea is by putting all the sentences together and figuring out what they are about. Read the following paragraph.

> Astronomers cannot do their job unless they have **telescopes** (tehl-uh-SKOHPS). Without telescopes, they cannot see the details of other planets or stars. With a telescope, astronomers can see both the color and the movement of stars. They can also see features such as mountains and valleys on planets.

Which sentence sums up the ideas in the paragraph?

___ A. By using a telescope, astronomers can see mountains on planets.

___ B. Telescopes help astronomers see the movements of stars.

___ C. Telescopes help astronomers see details of stars and planets.

The correct answer is C. A and B tell only one thing that telescopes help astronomers do.

Try It

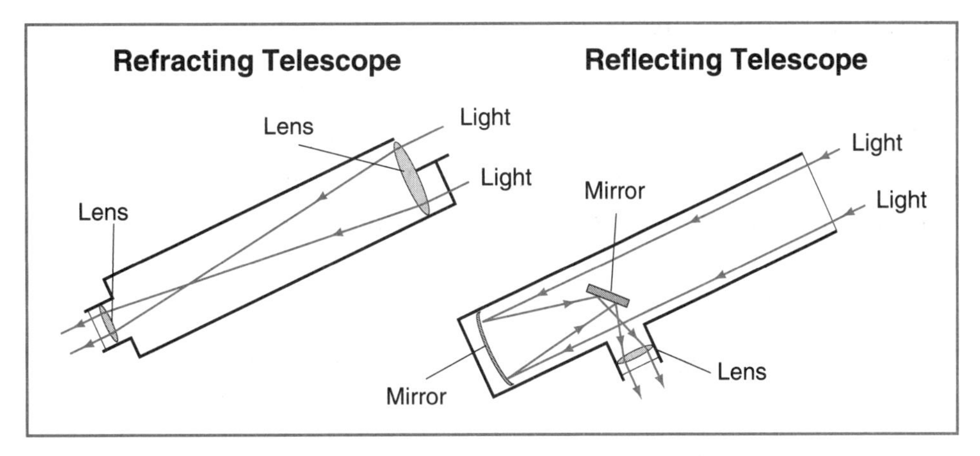

A refracting and a reflecting telescope are shown here.

Read the following paragraph. Then put a check mark next to the sentence that sums up the idea of the paragraph.

> A **refracting** (rih-FRAK-tihng) **telescope** is smaller than a **reflecting** (rih-FLEK-tihng) **telescope**. The refracting telescope uses two pieces of glass, called **lenses** (LEHN-zihz) to make something that is far away look bigger. The reflecting telescope uses two mirrors for the same purpose, but it does not show things as clearly.

___ **1.** Refracting telescopes are smaller than reflecting telescopes and show objects more clearly.

___ **2.** Reflecting telescopes use mirrors.

___ **3.** Refracting telescopes use lenses.

The correct answer is 1. The paragraph compares refracting telescopes and reflecting telescopes and tells how each works. Answers 2 and 3 are true, but they describe details, not the main idea.

Practice

Put a check mark next to the sentence that sums up the idea of each paragraph below.

1. In 1990 the United States sent a telescope into space. The Hubble Space Telescope travels around the Earth, but it is not pointed at the Earth. It is pointed toward outer space.

______ A. The United States has a space telescope.

______ B. The Hubble Space Telescope was launched in 1990 and is not pointed at the Earth.

______ C. The Hubble Space Telescope travels around the Earth and is used to explore outer space.

2. No clouds get in the way of the Hubble Space Telescope. There is no air pollution to block its view. It doesn't have to be shut down during bad weather. In addition, the Hubble Space Telescope can see seven times deeper into space than any telescope on Earth can.

______ A. Air pollution does not affect the Hubble Space Telescope.

______ B. The Hubble Space Telescope works better than any telescope on Earth.

______ C. The Hubble Space Telescope can see farther than telescopes on Earth.

Check your answers on page 119.

Follow-Up

Do you or does someone you know wear glasses? What happens when you are not wearing your glasses? Write a paragraph explaining how glasses and a telescope are similar.

Comparing the Planets

What You Know It's easy to see whether one line on a piece of paper is longer than another. If each line stands for a number you can tell which amount or number is bigger.

Look at the picture below. This kind of picture is often used to compare numbers. You can easily see which number is the largest by finding the longest line. You can also tell whether the second number is larger or smaller than the third.

How It Works

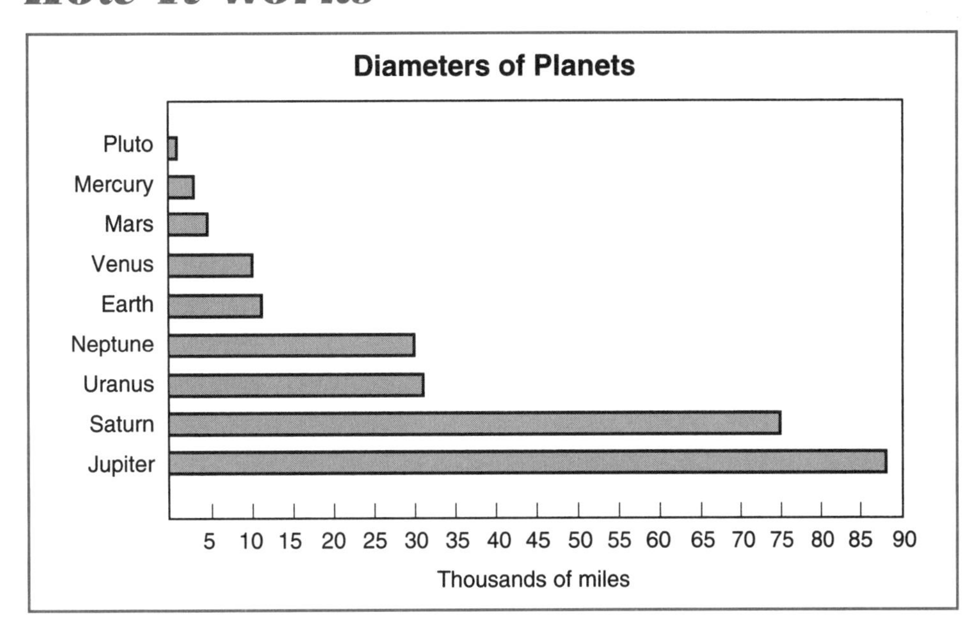

This bar graph compares the diameters of the planets.

This kind of graph is called a **bar graph**. First look at the title. It tells you what the graph is about. The title is "Diameters of Planets." A **diameter** (deye-AM-uht-uhr) is the length of a straight line that passes through the center of something. So this chart tells you the diameters of the nine planets.

Next look at the words on the left side of the graph. The names of the nine planets are shown there. Then look at the numbers that go across the bottom of the graph. These numbers are 5 through 90. Below the numbers it says "Thousands of miles" This means that 5 on the graph stands for 5,000 miles and that 75 on the graph stands for 75,000 miles.

Finally, look at the bars that go across the graph, starting at the left. Each bar shows how large the diameter of one of the planets is. See how long each bar is. Look from the far right end of the bar down to the numbers at the bottom. This gives you the diameter of the planet. For example, look at the names at the left and find Saturn. Then follow the bar for Saturn across until it ends. Then look down to find the number closest to where the bar ends. It is closest to 75 (or 75,000 miles), so Saturn is about 75,000 miles in diameter.

Try It

Look at the bar graph on page 97. Answer these questions. Place a check mark on the blank next to the correct answer.

1. Which planet is the smallest?

_____ Mercury

_____ Pluto

_____ Mars

The smallest planet is Pluto. The bar for Pluto doesn't reach past the number 5. It's closer to zero than to 5. This means that Pluto is about 1,000 to 2,000 miles in diameter. This diameter is smaller than that of any other planet on the graph.

2. Which two planets are about 30,000 miles in diameter?

_____ Mercury and Mars

_____ Neptune and Uranus

_____ Uranus and Saturn

Neptune and Uranus are both about 30,000 miles in diameter. Mercury and Mars are both less than 5,000 miles in diameter. Uranus is about 32,000 miles in diameter, but Saturn is about 75,000 miles in diameter.

Practice

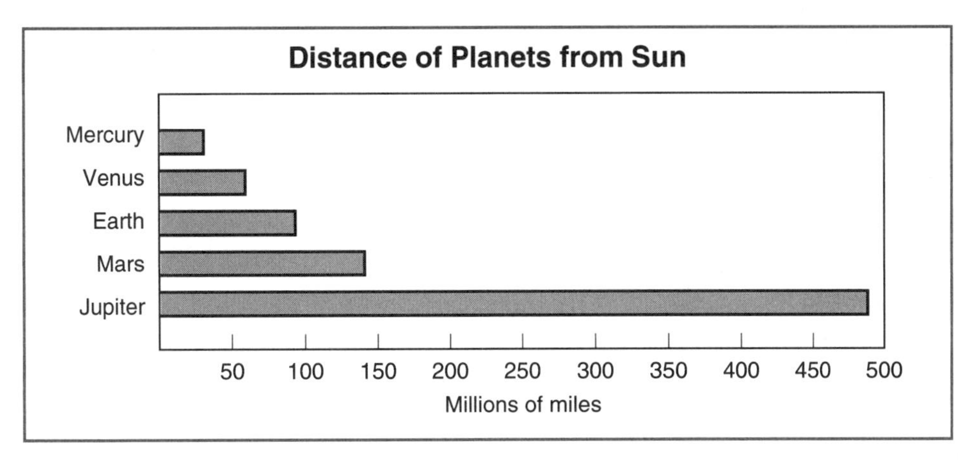

How far are the planets from the sun?

Look at the graph above. Put a check mark next to the correct answer.

1. Which planet is closest to the sun?

_______ A. Mercury

_______ B. Earth

_______ C. Jupiter

2. Which planet is about 480 million miles from the sun?

_______ A. Venus

_______ B. Mars

_______ C. Jupiter

Check your answers on page 119.

Follow-Up

Did you notice that only five of the nine planets were shown on the graph above? Do you know how far each of the four other planets is from the sun? (Saturn is 1.5 billion miles, Uranus is about 2.8 billion miles, Neptune is 4.5 billion miles, and Pluto is 5.9 billion miles, from the sun.)

Write two or three sentences explaining why you think that these four planets were not included on the graph.

Unit Reviews

Unit 1
Lesson 1 Review

Copy the circled words onto the long answer blanks below. Look at the first vowel in each word. On the short answer blanks, write S after words with short vowels and L after words with long vowels.

I (like) to (run) a (lot.)

It (takes) me (ten) minutes to (jog) a mile.

Sometimes my (wife) goes (with) me.

1. ______________ ______

2. ______________ ______

3. ______________ ______

4. ______________ ______

5. ______________ ______

6. ______________ ______

7. ______________ ______

8. ______________ ______

Lesson 2 Review

Put a check mark in front of the words that start with consonant blends.

1. ______ crab

2. ______ snug

3. ______ five

4. ______ trade

5. ______ lime

6. ______ fake

7. ______ trap

8. ______ slim

9. ______ rule

10. ______ bone

11. ______ spin

12. ______ drug

13. ______ save

14. ______ swam

Lesson 3 Review

Put a check mark next to the correct answer.

1. If you cook a hamburger too much, you
   ______ A. overcook it.
   ______ B. transcook it.
   ______ C. supercook it.

2. If something is not possible, it is
   ______ A. transpossible.
   ______ B. impossible.
   ______ C. superpossible.

3. If you move a bush in a garden, you
   ______ A. overplant it.
   ______ B. superplant it.
   ______ C. transplant it.

Lesson 4 Review

Write S on the answer blank in front of each pair of synonyms.
Write A on the answer blank in front of each pair of antonyms.

______ **1.** full empty

______ **2.** sleepy tired

______ **3.** joy happiness

______ **4.** thick thin

______ **5.** tall high

______ **6.** mist fog

______ **7.** off on

______ **8.** buy sell

Lesson 5 Review

Put a check mark in front of the words that end with consonant blends.

1. ______ gulf
2. ______ list
3. ______ slide
4. ______ tusk
5. ______ bond
6. ______ must
7. ______ book
8. ______ link
9. ______ mast
10. ______ lent
11. ______ meal
12. ______ frame
13. ______ gift
14. ______ blot

Lesson 6 Review

Look at the "Ten Tall Mountains" graph on page 17. Put a check mark next to the correct answer for each question.

1. How tall is Mount Everest?

    ______ A. About 3,000 feet
    ______ B. About 13,000 feet
    ______ C. About 30,000 feet

2. Which mountain is just over 20,000 feet tall?

    ______ A. Chimborazo
    ______ B. Mount Fuji
    ______ C. Broad Peak

Answers for Unit 1 Review begin on page 119.

Unit 2
Lesson 7 Review

Look at the underlined words in these sentences. Write the words with short vowels in the answer blanks under the word *Short*. Write the words with long vowels under the word *Long*.

Pat will take a jet to Los Angeles.

I hope she has time to visit Hollywood.

Short	Long
____________	____________
____________	____________
____________	____________
____________	

Lesson 8 Review

Put a check mark next to the correct answer below.

1. Ben is 5 feet tall, Jerry is 5 feet 5 inches tall, and Al is 6 feet tall.

______ A. Ben is the shorter.
______ B. Ben is the shortest.
______ C. Ben is the shortist.

2. A person who believes in social change is a

______ A. socialist.
______ B. socialest.
______ C. socialer.

3. The shirt costs $15, and the pants are $30.

______ A. The shirt is cheapist.
______ B. The shirt is cheapest.
______ C. The shirt is cheaper.

Lesson 9 Review

Read the sentences below. Then figure out what the underlined word in each sentence means. Place a check mark next to the correct answer.

1. Things you eat and drink can cause insomnia (ihn-SAHM-nee-uh). Doctors suggest that you not drink coffee or eat chocolate for at least six hours before going to bed. Both chocolate and coffee contain caffeine (kaf-EEN), which makes it hard for many people to sleep.

_____ A. an upset stomach

_____ B. the inability to sleep

_____ C. a headache

2. Eating too much can cause obesity (oh-BEES-uh-tee). When a person becomes very fat, the extra body weight can cause health problems.

_____ A. fatness

_____ B. headaches

_____ C. thinness

Lesson 10 Review

Read these sentences and then answer the questions.

The circulatory system is made up of the heart, blood, and blood vessels. Blood vessels are tiny tubes that carry blood to all parts of the body. There are about 60,000 miles of blood vessels inside your body. You heart beats about 100,000 times a day.

1. What are the parts of the circulatory system?

____

2. How many miles of blood vessels are there in your body?

____

3. About how many times does your heart beat every day?

____

Lesson 11 Review

Read the description on page 33 of how to help someone who is bleeding. Write the answers to the questions below on the answer blanks.

1. In step 1, what does *sit or lie down* mean?

____

2. What is the second thing you should do?

____

3. In step 4, what are *fresh cloths*?

____

4. Describe step 5 in your own words.

____

____

Lesson 12 Review

Look at the map that shows at-risk animals in the United States. It is on page 36. Put a check mark next to the correct answers below.

1. Where does the Jaguarundi live?

______ A. New York

______ B. Utah

______ C. Texas

2. Which state on this map contains no at-risk animals?

______ A. Arizona

______ B. North Carolina

______ C. Illinois

Answers for Unit 2 Review begin on page 120.

Unit 3
Lesson 13 Review

Read this paragraph and then answer the questions.

During the early 1970s, a group of people who worked on sound equipment tried a new type of sound recording. They recorded music using four microphones in four different places. The listener had to have four separate speakers to play back these recordings. Because most people didn't want to buy four speakers, **quadraphonic** (kwah-druh-FAHN-ihk) **sound** is not popular today.

1. When was quadraphonic sound invented?

____

2. Why isn't quadraphonic sound popular today?

____

Lesson 14 Review

Look at the following sentences. Write F in the answer blank if the sentence is a fact. Write O if the sentence is an opinion.

_____ **1.** The light in here is very bright.

_____ **2.** I put a 100-watt bulb in this lamp.

_____ **3.** Some 60-watt bulbs cost more than some 75-watt bulbs.

_____ **4.** I hate fluorescent lighting.

_____ **5.** This fluorescent lamp costs only $29.95.

_____ **6.** There are seven lights in the ceiling of this room.

Lesson 15 Review

Look at the lists below. Put a check mark next to the main idea of each list.

1. ______ A. Spanish

______ B. Japanese

______ C. Languages

______ D. French

2. ______ A. Florida

______ B. States

______ C. Washington

______ D. Mississippi

Lesson 16 Review

Look at the lists below. Show what you would do first, second, third, and fourth by writing the numbers 1, 2, 3, and 4 next to each step.

1. ______ Put one-third cup of uncooked oatmeal in a pan.

______ Serve the oatmeal in a bowl.

______ Add a cup of water to the uncooked oatmeal.

______ Cook the oatmeal for 5 minutes.

2. ______ Wipe the window dry with a clean cloth.

______ Use a cloth to apply the mixture to the window.

______ Fill a pail with a gallon of warm water.

______ Mix the juice of one lemon into the water.

Lesson 17 Review

Put a check mark next to the cause that most likely leads to the effect.

1. **Effect:** Shareen stopped coughing.

   Cause: ______ A. Shareen eats a lot of fried chicken.

   ______ B. Shareen stopped smoking.

   ______ C. Shareen's birthday was last week.

2. **Effect:** Ahmet learned how to swim.

   Cause: ______ A. He learned how to ride a bicycle.

   ______ B. His sister is 7 years old.

   ______ C. He took swimming lessons at the YMCA.

Lesson 18 Review

This drawing shows the convection and conduction of heat.

Look at the diagram above. Put a check mark next to the correct answers below.

1. What do the arrows going through the side of the cup mean?

   ______ A. The outside of the cup is cold.

   ______ B. Heat is passing through the cup.

   ______ C. Heat is staying in the cup.

2. What do the arrows pointing in the air mean?

   ______ A. Heat is rising from the cup.

   ______ B. Cold air is rising from the cup.

   ______ C. The cup is rising.

Answers for Unit 3 Review begin on page 121.

Unit 4
Lesson 19 Review

Read the following sentences. Write F in the answer blank if the sentence is a fact. Write O if the sentence is an opinion.

______ **1.** There is an electric power plant at Niagara Falls, New York.

______ **2.** This plant generates a lot of electricity.

______ **3.** The falls are over 150 feet high.

______ **4.** In 1757 Daniel Chabert Joncaire built a water-powered sawmill at Niagara Falls.

______ **5.** Niagara Falls is the most beautiful waterfall in the world.

Lesson 20 Review

Read the groups of sentences below. Put a check mark next to the main idea in each group.

1. ______ A. A single bus can carry 60 people.
______ B. Groups of people can carpool to work together.
______ C. There are many ways to save gasoline.
______ D. Electric trains are fast and cheap.

2. ______ A. A little thought can save money and energy.
______ B. Don't put a half-load of laundry in the washing machine. Wait until you have a full load.
______ C. Don't leave the TV on if you're not watching it.
______ D. Stuff cloth in cracks around windows to keep cold air from coming in during the winter.

Lesson 21 Review

Place a check mark next to the effect that most likely resulted from the cause.

1. Cause: Eric is working at two jobs.
Effect:
_____ A. He is less tired than he used to be.
_____ B. He is making twice as much money as he used to make.
_____ C. He is over 6 feet tall.

2. Cause: Alma fell and broke her arm.
Effect:
_____ A. She decided to get a job as a waitress.
_____ B. She went to the movies afterward.
_____ C. She stayed home for a few days.

Lesson 22 Review

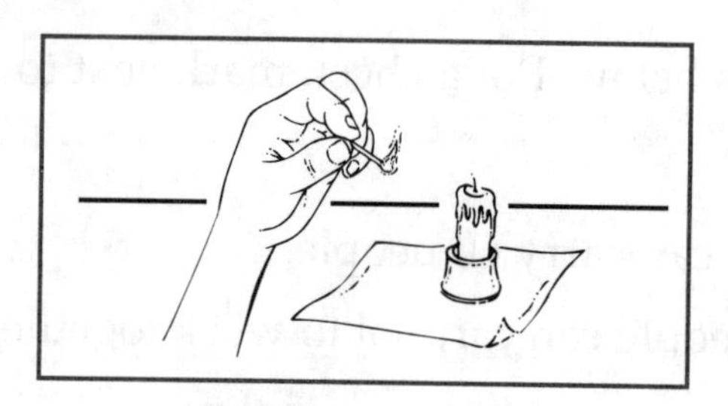

This is a drawing of a match, a candle, and paper.

Look at the picture above. Then put a check mark next to the correct answers below.

1. _____ A. The match has kinetic energy, but the candle doesn't.
_____ B. The candle has kinetic energy.
_____ C. Both the match and the candle have kinetic energy.

2. _____ A. The paper has potential energy, but the candle doesn't.
_____ B. Both the candle and the paper have potential energy.
_____ C. Both the candle and the paper have kinetic energy.

Lesson 23 Review

Read the following information. Then read the inferences below. Write Yes in the answer blank if you can make the inference from the information given and No if you cannot.

About 85 to 90 percent of the heat from the burning wood in a fireplace goes up the chimney. In spite of this, many modern houses have fireplaces. Fireplaces that put more of the heat into a room can be built, but they cost several hundred dollars more than a simple fireplace.

1. ______ Most of the heat from wood burning in a fireplace goes up the chimney.

2. ______ Most fireplaces are built of bricks.

3. ______ People don't want to pay several hundred dollars extra for a fireplace that puts more heat into the room.

4. ______ The government will pass a law against fireplaces.

Lesson 24 Review

Look at the "World Energy Use" pie chart on page 78. Put a check mark next to the correct answer to each question.

1. What percentage of the total amount of fuel used is gas?

______ A. 23.6 percent

______ B. 32.4 percent

______ C. 39.2 percent

2. What type of fuel accounts for 32.4 percent of the total fuel used?

______ A. Electricity

______ B. Gas

______ C. Solid fuels

Answers for Unit 4 Review begin on page 121.

Unit 5

Lesson 25 Review

Read the following sentences. Put a check mark next to the correct answer to each question.

The earth does not orbit (AWR-biht) the sun in a perfect circle. At certain times of the year, the earth is closer to the sun than at other times. The earth makes an elliptical (eh-LIHP-tih-kuhl) path around the sun. This path looks like a circle that has been stretched out on two opposite sides.

1. What does *orbit* mean?

______ A. block out

______ B. move around

______ C. explode

2. What does *elliptical* mean?

______ A. not quite round

______ B. very fast

______ C. dangerous

Lesson 26 Review

Underline the sentence that contains the main idea in this paragraph.

A monsoon is a pattern of wind in parts of India and Southeast Asia. From April to October, the monsoon blows from the southwest, bringing heavy rains. From November to March, it blows from the northwest, drying out the land. It mainly affects the eastern coast of India. However, it also affects such countries as Laos, Thailand, and Vietnam. The heaviest rainfall occurs at Cherrapunji, in northeastern India, where as much as 1,042 inches of rain have fallen in one year.

Lesson 27 Review

Read this paragraph about Earth and Mars. Then put a check mark next to the correct answer to each question below.

> Earth and Mars may seem very different, but they do have some things in common. For example, scientists have discovered ice, but no water, on Mars. We also know that a day on Mars has the same number of hours as a day on Earth. (A day on Venus lasts as long as 243 Earth days.) Mars has two moons, and its year is 687 days long. This means that Mars takes almost twice as long to travel around the sun as the Earth does.

1. How are Earth and Mars the same?

______ A. A year is the same length on both planets.

______ B. They both some ice on their surfaces.

______ C. They both have one moon.

2. How are Earth and Mars different?

______ A. A day on Mars is twice as long as a day on Earth.

______ B. A year on Mars is shorter than a year on Earth.

______ C. Mars has two moons.

Lesson 28 Review

Reread the paragraph in Review Lesson 27. Then write Yes on the answer blank if you have enough information to make the inference. Write No if you do not.

______ **1.** A day on Mars lasts 24 hours.

______ **2.** Both moons of Mars are larger than Earth's one moon.

______ **3.** Some parts of the surface of Mars are very cold.

Lesson 29 Review

Put a check mark next to the idea that sums up the main idea of the paragraph below.

> **Binoculars** (beye-NAHK-yuh-luhrz), like telescopes, make distant objects appear nearer and clearer. Using binoculars is like having a separate telescope for each eye. Binoculars allow users to see a wider area than they can with a telescope. Also, if one eye is stronger than the other, the user can make adjustments to the binoculars. Binoculars also make it easy to tell how far away objects are. They are also lighter and smaller than a telescope. They can even be made small enough to carry in a pocket.

_____ A. Binoculars make distant objects appear larger.

_____ B. Binoculars help the user see a wide area.

_____ C. Binoculars have many advantages over telescopes.

Lesson 30 Review

Look at the bar graphs on pages 97 and 99 and answer these questions.

1. Which is the second-smallest planet?

_____ A. Jupiter

_____ B. Pluto

_____ C. Mercury

2. Which planet is about 93 million miles from the sun?

_____ A. Earth

_____ B. Mars

_____ C. Venus

3. Which planet is just over 50 million miles from the sun?

_____ A. Venus

_____ B. Earth

_____ C. Jupiter

Answers for Unit 5 Review begin on page 122.

Answers

Unit One

LESSON 1

Practice

1. space/long; gas/short
2. dust/short; like/long
3. spin/short; fast/short
4. came/long; be/long

LESSON 2

Practice

1. skin, skin, crust
2. plate, broken, crust
3. plates, floor
4. States, plate, slowly, floating
5. plates, crash
6. plates, slide

LESSON 3

Practice

1. impossible; *im* means *not*
2. transported; *trans* means *across*

LESSON 4

Practice

1. A. small
1. B. large
2. A. chilly
2. B. warm

LESSON 5

Practice

1. important, effect
2. warm, current, upward
3. cold, wind, downward
4. act

LESSON 6

Practice

1. B
2. B
3. A
4. A

Unit Two

LESSON 7

Practice

1. plant/short; life/long
2. use/long; make/long
3. like/long; can/short
4. cells/short; big/short
5. little/short

LESSON 8

Practice

1. harder
2. artist
3. safer
4. faster
5. fastest

LESSON 9

Practice

1. Resistance is the ability to fight off diseases before they start.
2. Aerobics is a kind of exercise.

LESSON 10

Practice

1. A baby's heart begins to beat several weeks after it is conceived.
2. Blood enters the heart on the right side.
3. It pumps the blood into the left side of the heart.
4. The left side of the heart pumps the blood out to the rest of the body.

LESSON 11

Practice

1. A. When one foot is *ahead* of the other, one foot is a bit in front of the other.
2. B. Step 5 says to use your leg muscles to lift the box.

LESSON 12

Practice

1. C
2. B
3. A
4. C

Unit Three

LESSON 13

Practice

1. Thomas Edison
2. in 1877
3. He spoke into a large cone

LESSON 14

Practice

1. F; you can see it.
2. O; notice the word *nice.*
3. F; you can see it.
4. O; notice the words *I don't like.*
5. O; there may be different opinions about how popular they are.

LESSON 15

Practice

1. height
2. time
3. liquid amounts
4. numbers

LESSON 16

Practice 1

The answers are 5, 3, 1, 2, and 4.

Practice 2

The answers are 4, 5, 2, 1, and 3.

LESSON 17

Practice

1. The third answer is correct. The deer makes the stone move.
2. The first answer is correct. Inertia keeps the car moving.
3. The second answer is correct. Inertia keeps the dresser from moving.

LESSON 18

Practice

1. Convection of Heat
2. teakettle, burner
3. water, teakettle, burner
4. The arrows show the direction the water moves inside the teakettle as it gets hot.
5. First the water at the bottom gets hot and rises. Then the cool water at the top moves to the bottom.
6. This movement of heat through a liquid is called convection.

Unit Four

LESSON 19

Practice

1. F; you can count the amount of money.
2. F; you can count the amount of money.
3. O; notice the words *I don't think.*
4. F; you can measure the amount of electricity produced.
5. O; notice the word *best.*
6. F; you can prove that this commission controls the operation of nuclear plants.
7. O; this is how someone feels.

LESSON 20

Practice

1. Adobe houses never get too hot or too cold inside.
2. Adobe bricks are cheap and easy to make.

LESSON 21

Practice

1. The second answer is correct. Black glass is darker than clear or yellow glass.
2. The first answer is correct. Solar collectors face south to get the hottest rays of the sun.
3. The third answer is correct. Solar collectors do not work on cloudy days or at night. This means that they don't give a steady supply of energy.

LESSON 22

Practice

1. The second answer is correct. The first and third answers are wrong because the firecracker doesn't have kinetic energy. It has potential energy.
2. The third answer is correct. The first and second answers are wrong because the firecracker contains potential energy, not kinetic energy.

LESSON 23

Practice

1. Yes
2. No
3. No
4. Yes
5. No

LESSON 24

Practice

1. A
2. C
3. C

Unit Five

LESSON 25

Practice

1. C
2. A

LESSON 26

Practice

1. Seasons change because each part of the Earth receives different amounts of the sun's energy at different times of the year.
2. Some parts of the Earth don't have seasons.

LESSON 27

Practice

1. B. A is not true—Mars does not have a ring. C is not true—Jupiter is the largest planet.
2. C. A is not true; all the planets are round. B is not true because both Mars and Jupiter have more than one moon.

LESSON 28

Practice

1. No. The stars look the same in different countries around the world. The paragraph doesn't mention stars looking different in different places.
2. Yes. The three groups of people mentioned in the paragraph see three different things in the same group of stars.
3. No. The paragraph doesn't mention people in Spain.
4. No. The paragraph doesn't mention why different people see different things.

LESSON 29

Practice

1. C. A and B are details from the paragraph.
2. B. A and C are true, but they are details from the paragraph, not the main idea.

LESSON 30

Practice

1. A
2. C

Unit One Review

LESSON 1

1. like, L
2. run, S
3. lot, S
4. takes, L
5. ten, S
6. jog, S
7. wife, L
8. with, S

LESSON 2

These numbers should be checked: 1, 2, 4, 7, 8, 11, 12, and 14.

LESSON 3

1. A
2. B
3. C

LESSON 4

1. A	**2.** S	**3.** S	**4.** A
5. S	**6.** S	**7.** A	**8.** A

LESSON 5

These numbers should be checked: 1, 2, 4, 5, 6, 8, 9, 10, and 13.

LESSON 6

1. C **2.** A

Unit Two Review

LESSON 7

Short: Pat, will, jet, has
Long: take, hope, time

LESSON 8

1. B **2.** A **3.** C

LESSON 9

1. B **2.** A

LESSON 10

1. the heart, blood, and blood vessels
2. 60,000 miles
3. about 100,000 times

LESSON 11

1. It means that the person who is bleeding should sit down or lie on his or her back.
2. Press a clean cloth to, or wrap a cloth around, the bleeding area.
3. *A fresh cloth* is a clean cloth, without blood on it.
4. Answers will vary. One possibility is to telephone a hospital or doctor right away.

LESSON 12

1. C **2.** C

Unit Three Review

LESSON 13

1. during the early 1970s
2. because people don't want to buy four separate speakers

LESSON 14

1. O
2. F
3. F
4. O
5. F
6. F

LESSON 15

1. C
2. B

LESSON 16

1. The order is: 1, 4, 2, 3
2. The order is 4, 3, 1, 2.

LESSON 17

1. B
2. C

LESSON 18

1. B
2. A

Unit Four Review

LESSON 19

1. F
2. O
3. F
4. F
5. O

LESSON 20

1. C
2. A

LESSON 21

1. B
2. C

LESSON 22

1. A
2. B

LESSON 23

1. Yes **2.** No **3.** Yes **4.** No

LESSON 24

1. A **2.** C

Unit Five Review

LESSON 25

1. B **2.** A

LESSON 26

The main idea is found in the first sentence: A monsoon is a pattern of wind in parts of India and Southeast Asia.

LESSON 27

1. B **2.** C

LESSON 28

1. Yes **2.** No **3.** Yes

LESSON 29

C

LESSON 30

1. C **2.** A **3.** A